Basic Electronics Engineering

Satya Sai Srikant · Prakash Kumar Chaturvedi

Basic Electronics Engineering

Including Laboratory Manual

 Springer

Satya Sai Srikant
Department of Electronics &
Communication Engineering
SRM Institute of Science and Technology
Ghaziabad, Uttar Pradesh, India

Prakash Kumar Chaturvedi
Formerly in Department of Electronics &
Communication Engineering
SRM Institute of Science and Technology
Ghaziabad, Uttar Pradesh, India

ISBN 978-981-13-7416-6   ISBN 978-981-13-7414-2   (eBook)
https://doi.org/10.1007/978-981-13-7414-2

This Springer imprint is published by the registered company Springer Nature Singapore Pte Ltd.
The registered company address is: 152 Beach Road, #21-01/04 Gateway East, Singapore 189721, Singapore

Preface

Basic Electronics Engineering is an elementary course compulsory for engineering students of all disciplines. We hope to bridge the gap in basic electronics engineering by providing a textbook which matches syllabus compatibility with right pedagogy. This book is an outcome of our vast experiences both in industries as well as in academics. This book is based on the syllabi of Basic Electronics of major Indian universities.

The book covers topics of basic electronics engineering ranging from basic electronics to electronic circuits to transducers to digital systems to communication systems. The chapters provide a comprehensive exposition of the principles of electronics engineering for all first year engineering students who just came out from elementary science courses. The language used in this book is simple and lucid to impart knowledge to all level of students. For better visualization of concepts, a number of self-explanatory figures have been used. Each topic has been explained systematically and analytically. This book can also be used for the preparation of competitive exams.

The book consists of eight chapters followed by Annexure giving related constants and finally Index. Chapter 1 introduces the subject along with the concept of band theory, metal semi-conductors and insulators. Chapter 2 explains the diodes and its applications as rectifier, clippers, etc. Chapter 3 gives the types of transistors, e.g., Bipolar Transistors, Unipolar Transistors like JnFET, MOSFET, their constructions, working and comparisons. Chapter 4 introduces the optoelectronic devices, e.g., LED, LDR, Solar cell, Photodiodes, Photo-transistors, opto-couplers, LCD and Infrared emitters. Chapter 5 deals with digital electronics and discusses number systems and codes, followed by Boolean algebra, logic gates and Karnaugh map. Chapter 6 deals with various types of transducers with its applications. Chapter 7 deals with communication systems and throws light on the concept of waves, electronic signals, various modulation techniques and applications in transmitter/receivers, satellite communication, RADAR, TV, radio, etc. Chapter 8 gives twelve most important basic lab experiments performed in academic environments for engineering students as well as faculty members.

We are especially thankful to our partners and children who have been a source of inspiration and provided us with support even in odd hours. We are also thankful to the authors of various books we have referred during teaching sessions as well as while writing this book.

Ghaziabad, India   Dr. Satya Sai Srikant
                                                                        Prof. Prakash Kumar Chaturvedi

About This Book

The *Basic Electronics Engineering Including Laboratory Manual* is normally a first course of Electronics, which is compulsory for engineering students of all disciplines, who are perusing for their B.Tech./B.E. Degree courses. Since, many engineering students find it very difficult to conceptualize the subject; therefore it has been presented in a friendly and lucid language. In its eight chapters there has been an attempt to prepare the engineering students for an in-depth study and is an outcome of our experiences in industries as well as in academics.

Key Features

- Easy explanation of the topics, plenty of practical examples and illustrations.
- The text is supplemented by a very large number diagrams (over 200), with special effort put in for giving numerical values in graphs, tables etc. in order to get a real feel/visualization of the device under study, for better understanding of the subject.
- Special effort has been to give actual numerical values in the scales of the graphs, dimensions of the components/devices etc., for getting a real visualisation of the device and their properties, as this is missing in most of the books in this subject.
- Description details of the figures/graphs are given there itself in addition to that in the text material.
- Large number of solved problems, long/short review question, fill in the blanks etc. taken from Examination papers of various universities, for the practice of the students
- Additional feature is the chapter on "Basic Lab experiments with lab manual" along with viva/quiz questions for the benefit of the students and faculty members
- As standard book an appendix giving all the necessary constants has been given followed by references and index.

Contents

About the Author

Dr. **Satya Sai Srikant** is an Associate Professor at SRM Institute of Science and Technology, Modinagar, UP since 2009. He pursued his M.Tech. from the University of Delhi in Microwave Electronics in 2002 and Ph.D. from Siksha 'O' Anusandhan University, Bhubaneswar in 2014. His research contributions in the field of application of microwaves in minerals and materials processing in RF and Microwave industries have included the design of oscillator, amplifier, filters and antennas. He has published more than 40 papers in reputed journals, conferences and seminars. He teaches basic electronics, RF transmission lines, signals and system, microwave communication, electromagnetic fields and waves to B.Tech. and M.Tech. engineering students. He is presently guiding Ph.D. students and working on research projects in the area of microwaves.

Prof. Prakash Kumar Chaturvedi was formerly a Professor at SRM Institute of Science and Technology, Modinagar, UP. After his M.Tech. from BITS Pilani in 1969, he pursued Ph.D. from CEERI, Pilani in 1974 and a MBA from University of Stirling, Scotland. He has been Project Management Chief of several Government of India projects like digital TV, CODIN, Technology Development for Indian Languages (TDIL), etc. He has guided several Ph.D. students and has a rich experience in teaching and research. He is also associated with the Entrepreneurship Development Program at SRM University.

Chapter 1
Semiconductor—An Overview

Contents

© Springer Nature Singapore Pte Ltd. 2020
S. S. Srikant and P. K. Chaturvedi, *Basic Electronics Engineering*,
https://doi.org/10.1007/978-981-13-7414-2_1

1.1   Introduction

Whether it is electrical system or electronics, both deal with electrical parameters like charge, current, voltage, frequency resistance, inductance, capacitance, etc. Therefore, there is a lot of overlap between the two areas; still the differences between the two are very much clear in terms of the following three things:

(a) **Levels of Power, Current and Voltage**: In electronics, the power may typically be <10 W, current <1 A and voltage <30 V.

(b) **Applications**: As far as applications are concerned, the electrical engineering deals with generation, distribution of power and utilization of power, while in electronics, we deal with information, signals, their generation, receiving, transmission, transforming, processing, storing, coding, etc.

(c) **Size of Devices**: The sizes of electrical components like generators, motors, etc. are very large as compared to electronics like transistors, IC, etc. Largest electronic devices may be microwave tubes, transmitters, receivers, Radars antennas, etc.

The growth of electronics was from the invention of telegraph in Germany in 1837 by Gauss and Hocher to T.V. in 1927, to semiconductors in 1931, to transistors by Shockley in 1948, to microprocessor in 1971, to satellite and mobile communication after 1990 and so on. Now whichever gadgets/appliance we use whether electrical or electronics or mechanical has semiconductor devices in it, e.g. diodes, transistor, IC, etc.

In this chapter, we will deal with semiconductor which can be classified as given below:

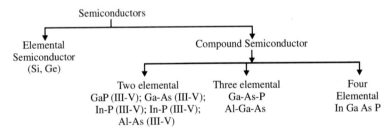

All the elemental semiconductors basically belong to Group IV elements, where as all the compound semiconductors are formed by the elements of Group III and V of the periodic table with the following applications:

Elemental and compound semiconductor	Applications
Si, Ge	All types of devices like diode, transistors IC, etc.
GaAs, GaP, GaN, $GaAs_{1-x}P_x$, GaTe, AlN, InGaN	LED[a]. The x value (0–1) in $GaAs_{1-x}P_x$ decides the band gap 1.4–2.7 eV, and hence colour of the LED changes from IR to UV; other compounds are also used for different colours
GaAs, In–P	Microwave devices
ZnS	Fluorescent materials for screen, tube light, etc.
GaAs, AlGa As	Lasers
In–Sb, Cd–Sa, P Li–Te	Light detectors
GaAsP, InGaAsP	Power electronics products

[a]*Note* Band gap E_g < 1.7 eV, corresponds to $\lambda \geq 0.73$ μm (infrared)
While E_g > 3.0 eV, Corresponds to $\lambda \geq 0.41$ μm (ultraviolet)
Here, $E_g = h\nu = h \cdot c/\lambda$; $\lambda = 1.24/E_g$ (with λ in μm and E_g in eV)

1.2   Structure of Solids

The study of X-ray scattering has revealed the fact that most of the solids are crystalline in structure. In other words, a solid consists of atoms or molecules which are arranged in a periodic manner. There is always some basic arrangement of atoms, which is repeated throughout the entire solid materials. Such an arrangement of atoms within a solid is called crystal lattice. Such solids are called crystalline solids. There are also some other solid materials which do not have crystalline structure. Such solid materials are called non-crystalline or amorphous solids. All metals and semiconductors like germanium, aluminium and silicon are crystalline materials, whereas non-metals like wood, plastic, glass, papers, etc. are amorphous (i.e. shapeless) solid materials. Figure 1.1 shows the structures of crystalline and amorphous solids.

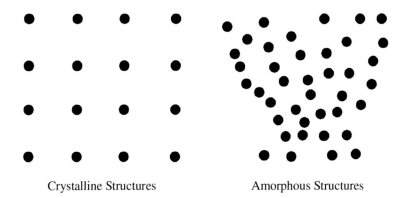

Crystalline Structures   Amorphous Structures

Fig. 1.1 Atoms in the structure of crystalline and amorphous solids

1.3 Atomic Structure and Energy Levels

The most fundamental unit of matter is an atom. On the basis of experiments of the scattering of α-particles, combined with the discovery of neutrons by Chadwick in 1932, Rutherford suggested a complete model of the atom which is known as *Rutherford's Nuclear Model of Atomic Structure*. According to his model (Fig. 1.2):

(a) An atom consists of a central portion which is called the nucleus. All the protons and neutrons are present in the nucleus. Although the nucleus is small, it is heavy due to the presence of all the protons and neutrons in it. Since the mass of the electron is negligible, the mass of an atom is equal to the sum of the masses of protons and neutrons, i.e. the entire mass of an atom resides in its nucleus.

(b) The electrons keep revolving in the orbits around the nucleus at extremely high speed and at large distance from the nucleus. It is similar to the planets of solar system, where the planets revolve around the sun as shown in Fig. 1.2. Therefore, the electrons revolving around the nucleus are sometimes called planetary electrons.

Since the atom is neutral, the number of electrons which are negatively charged particles is equal to the number of protons in the nucleus. The electrons move in approximately elliptical orbits about the nucleus.

Thus the three particles, electrons, protons and neutrons, are called the fundamental particles of the universe. It means that there is no difference between an electron in an atom of silver and an electron in an atom of carbon. Different elements behave differently because of the fact that there is a difference in the number and arrangement of electrons, protons and neutrons in atoms of different elements.

The number of protons or electrons in an atom is called its **atomic number**. For example, hydrogen (H) atom has one proton in its nucleus and one electron around, its nucleus. So, the atomic number of hydrogen atom is 1.

Fig. 1.2 Rutherford's nuclear model of atomic structure of Li

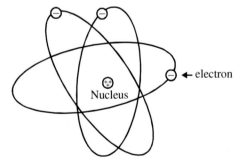

The atomic number and valency of some important elements used in semiconductors are given below:

	Trivalent				Tetravalent			Pentavalent		
Atom	B	Al	Ga	In	C	Si	Ge	P	As	Sb
Atomic number	5	13	31	49	6	14	32	15	35	51

All the electrons of an atom do not move in the same orbit. All the electrons are arranged in different orbits or shells. These shells may be circular or elliptical and are numbered as K, L, M, N, O, P and Q with K being nearest to the nucleus. Each shell can contain only a particular number of electrons. It was Bohr, who invented a model called Bohr's model indicates that maximum number of electron in a shell is $2n^2$; where n is the number of the shell or level. The first shell is K; second shell is M. K shell can have a maximum of two electrons. Similarly, other shells $L = 2 \times 2^2 = 8$, $M = 2 \times 3^2 = 18$, $N = 2 \times 4^2 = 32$ electrons.

In each of these orbits or shells also, there are sub-orbits (called orbitals) named as s, p, d, f which are given numbers $l = 0, 1, 2, 3$, respectively, and can accommodate a maximum of $(2 + 4l)$ numbers of electrons, i.e. $s = 2$, $p = 6$, $d = 10$, $f = 14$. Therefore, the electrons in orbits are divided into sub-orbits (i.e. orbitals) as $K = 2(s)$; $L = 8 = 2(s) + 6(p)$; $M = 18 = 2(s) + 6(p) + 10(d)$ and $N = 32 = 2$ $(s) + 6(p) + 10(d) + 14(f)$. Therefore, in silicon, the orbits and orbitals will be $K = 2(s), L = 8 = 2(s) + 6(p); M = 4 = 2(s) + 2(p)$, i.e. the outermost orbit M has two electrons each on s and p orbitals, with p-orbital remaining unfilled by four electrons, making silicon as tetravalent. Therefore, the electrons in the outermost orbit which determines the electrical and chemical characteristics of each particular type of atom are referred to as valence electrons. An atom may have its outer or valence orbit completely filled or only partially filled.

The atomic number of aluminium is 13. The aluminium has 13 protons and 14 neutrons in its nucleus. Bohr's model of aluminium atoms shows three orbits. The first orbit contains $2 \times 1^2 = 2$; the second orbit contains $2 \times 2^2 = 8$ and the third orbit contains remaining 3. So, there are three electrons (valence electrons) in the outermost shell of aluminium atom, which makes it trivalent, as shown in Fig. 1.3a. Similarly, Fig. 1.3b shows that there are four valence electrons for silicon atom,

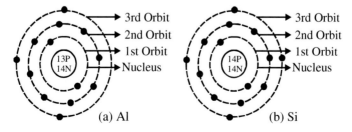

Fig. 1.3 a Atomic structure of Al (13). b Atomic structure of Si (14)

having 14 atomic number and having 14 protons and 14 neutrons in its nucleus. Hence, its valency = 4.

1.3.1 Energy Levels of Electrons in Isolated Atoms

As discussed above as per Bohr's model, the electrons can revolve only in the permitted orbits but not in any intermediate orbit. It means that each orbit has fixed amount of energy mv^2/r associated with size of its orbit. The larger the orbit, the greater is its energy. Hence, electrons in the outermost orbits have more energy of its own than the electrons in the inner orbit, and less external energy (heat or radiation) is required to make it free from the atom. However, the absolute energy (mv^2/r) of the innermost orbit is higher, requiring very large amount of energy to make it free to have zero energy. Figure 1.4a shows the energy level (orbit) diagram for silicon atom. The permitted orbits and then energy level are numbered as $n = 1, 2, 3, \ldots$ in increasing order of energy (when electron goes from excited higher orbit to lower, it releases (looses) energy of radiation, therefore lower orbit has lower/more -ve valued energy).

The level of energy obtainable with different orbits can be represented by horizontal lines, as shown in called **energy level diagram. Here, the free electron is shown to have zero energy**.

Thus, electrons closest to the nucleus need greatest energy (being bonded with neucli more strongly) for extracting them from the atom. The electron in inner orbits thus remains bound to the nucleus and therefore, called bound electrons.

The energy levels considered above are measured in electron volts (eV). An electron volt is defined as the energy required to move an electron through a potential difference of 1 V. In the unit of work or energy, the electron volt (eV) is given as

$$1\,eV = 1.602 \times 10^{-19}\,J$$

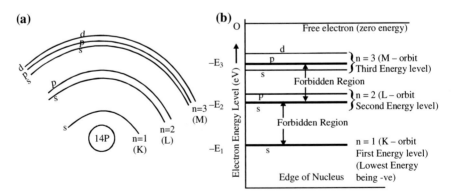

Fig. 1.4 Energy levels in isolated Si atom orbits (K, L, M) and their sub-orbitals, i.e. orbitals (s, p, d)

Between each energy level, a forbidden region exists. Electrons cannot orbit/revolve in forbidden regions. However, they may quickly pass then. If electrons in a particular orbit acquire energy from some outside source (such as radiation, external field or thermal energy), they will jump to a higher energy level by passing quickly through the forbidden region. A definite amount of energy is required to be spent in order to remove an electron from its orbit (level). This electron which has gone to high orbits are unstable, and they tend to fall back to low energy level by giving up their additional energy. The electrons that do leave their parent atom are termed as free electrons and said to have zero energy.

1.3.2   Interatomic Forces of Closer Atoms like in Solids

We have seen the energy levels of electrons in an isolated atom, but in real world this is rarely the case. In fact in solids (or even in liquid or gas), the atoms are quite close, where two types of forces (F_a and F_{es}) between the atoms come into play, as shown in Fig. 1.5.

 (i) **Atomic Forces of attraction (F_a):** This is due to co-valent bond, −ve and +ve ions etc., keeps the atom bonded.
 (ii) **Electrostatic forces of repulsion (F_{es}):** This is between orbits of electron of two atoms. It increases very rapidly and at $d = d_0 = 2.3$ Å, the electrostatic force of repulsion (F_{es}) is equal to atomic force of attraction (F_a), i.e. $F_{es} = F_a$. For $d < d_0$, F_{es} dominates. Thus, the attraction between two atoms becomes repulsive when they come closer than d_0 or so (This is true in human beings also, between two friends/relatives). These atoms then settle at this distance of $d = d_0$ to make crystals or solids.
 (iii) **Nuclear force (F_a):** These are forces between neutron to neutron (NN), proton to proton (PP) and neutron to proton (NP). These forces are effective at very short distances of $d \leq 10^{-15}$ m and diminish very fast with distance.

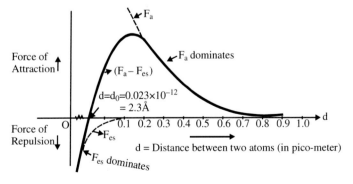

Fig. 1.5 Forces (F_a and F_{es}) between atoms when brought closer in picometer (10^{-12} m) range

1.4 Energy Levels of Electrons in Solids

When atoms come closer at a stable distance (d_0) to form gas, liquid or solids, their orbits influence each other and the allowed energy levels increase, i.e. each of the levels of $1s^1$, $2s^2$, $2p^1$, $3s^2$, $3p^2$ in silicon ($Z = 14$) split into a large number of levels (Fig. 1.6a). The inner orbitals also split but much less. The outer most orbitals ($3p^2$) and ($3s^2$) split maximum to cross each other and form into so closely packed levels or states that they form a continuous band of energy called conduction and valence band with a gap called band gap. It can be understood by a figure shown in Fig. 1.6b, in silicon for all the orbitals at the interatomic distance d_0.

The number of closely packed energy levels is the same (N) as the number of atoms (N) in the solid having 10^{23} atoms/cc. The electron in the outermost orbit has highest energy and by varying little additional energy (e.g. thermal or optical), it becomes free electrons (i.e. zero energy) that move freely for conduction in solid. The innermost orbital electrons (e.g. $1s^2$) have least energy (-1823 eV) of its own and require a very very large amount of energy for getting removed.

The $3p^6$ orbitals can have six electrons energy levels, but silicon has only two electrons in it ($3p^2$). When the conduction band is formed by crossing $3p^2$ and $3s^2$ orbitals, the allowed states in conduction band are 4N, with no electrons present there while the valence band has 4N states all filled with 4N electrons at 0 °C.

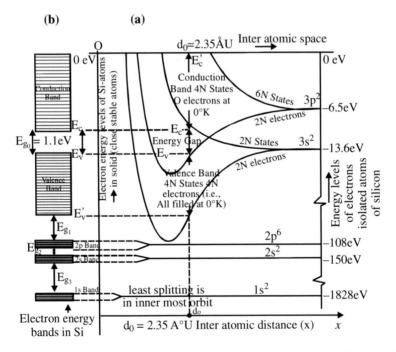

Fig. 1.6 a Splitting of energy electron energy, states/levels, their occupation and **b** energy band formation in silicon of all the orbitals

However, at a higher temperature, i.e. room temperature, due to greater thermal vibration of atoms, the bonding of electrons in the valence band is weak, so they jump to conduction band making the material a partial conductor called semiconductor.

1.5   Energy Band Structures

Figure 1.7a shows the energy band diagram of silicon for electrons of outermost orbitals ($3s$ and $3p$) only. The energy band that possesses the valence electrons is called valence band. The third orbit of an isolated silicon atom as shown in Fig. 1.3b is not completely filled, having only four electrons. However, in solid silicon, each atom is surrounded by four other silicon atoms and these four atoms share an electron with the central atom. Hence, each atom has eight electrons in a solid and therefore the valence band is completely filled. Due to four valence electrons, the silicon atom is a tetravalent element. So, silicon (Si) forms a tetrahedron structure and an atom in the centre of the tetrahedron shares electrons with atoms on each vertex. The atoms of silicon are arranged in an orderly pattern and form a crystalline structure as shown in Fig. 1.7b, whereas Fig. 1.7c shows a two-dimensional representation of such a structure as well as the covalent band structure of Si.

The energy band which possesses the free electron (which are thermally generated at room temperature) is called conduction band. Electrons in this conduction band take part in conduction of current. If a substance has empty conduction band, it means that current conduction is not possible in those substances, which is true, that is, 0 °K for all elements.

For germanium (Ge), the atomic structure is shown in Fig. 1.8a, while Fig. 1.8b shows its energy band diagram. Its atomic number is 32. So, it has 32 protons in the nucleus and 32 electrons distributed in the four orbits around the nucleus.

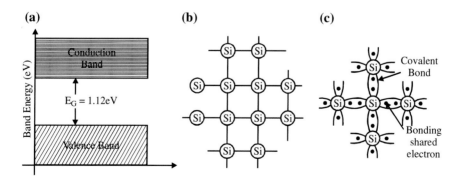

Fig. 1.7   a Energy band diagram of solid silicon, **b** crystalline structure of silicon, **c** covalent bond in Si

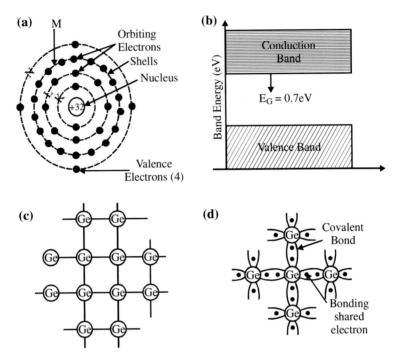

Fig. 1.8 **a** Atomic structure of Ge, **b** energy band diagram of Ge, **c** crystalline structure of Ge, **d** covalent bond for Ge

Similarly, the germanium atom has four valence (outermost orbit) electrons and is devoid of four electrons in its orbitals for saturating that orbit; hence, it is tetravalent element. The crystalline structure of Ge is shown in Fig. 1.8c, whereas covalent structure for Ge is shown in Fig. 1.8d.

1.5.1 Forbidden Energy Gaps and Band Structures

We know that between the valence band and conduction band, there is a forbidden energy gap (E_G). Energy is required to lift the electrons from valence band to the conduction band which has to be larger than the energy gap E_G. If we provide less energy than E_G, the electron will not get lifted because there are other allowed energy levels for electron to stay between valence band and conduction band. In other words, forbidden energy gap (E_G) is a region in which no electron can stay. To make more of the valence electron free, i.e. transfer to conduction band, some external energy is required, e.g. heat or light, which has to be at least equal to the forbidden energy gap. The forbidden energy gap (E_G) is 1.12 eV for silicon and 0.72 eV for germanium. At 0 °K, the number of electrons presents in the conduction band is zero; hence, conductivity is zero.

Semiconductor element	Ge	Si	GaAs	GN	SiC	InP	InAs	InSb	C
(Energy Gap E_G in eV at 300 °K)	0.72	1.12	1.42	3.44	2.36	1.3	0.35	0.17	5.5

At room temperature (300 °K), a minute quantity of valence electrons is lifted to conduction band due to thermal energy and this constitutes some current conduction, making it a semiconductor. However, at room temperature (300 °K), the number of electrons lifted to the conduction band in case of silicon is quite less than germanium. This is one of the reasons why silicon semiconductor devices are preferred over germanium devices, as the above reason leads to low leakage current (I_0) in Si diodes.

1.5.2   Insulators, Metals and Semiconductors on the Basis of Energy Band Gap

For any given material, the forbidden energy band gap (E_G) may be small, large or non-existent. Based on this band gap, the electrical properties of insulators, metals and semiconductors will be discussed now.

1.5.2.1   Insulators

In insulators even at room temperature (300 °K), there are generally no electrons in the conduction band, while the valence band is filled. In other words, valence band filled implies that all the permissible energy levels in this band are occupied by the electrons. The energy band diagram of insulators is shown in Fig. 1.9a. The energy

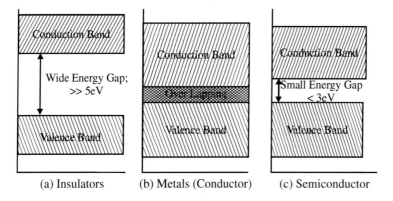

(a) Insulators   (b) Metals (Conductor)   (c) Semiconductor

Fig. 1.9 Energy band diagram of insulator, metal and semiconductor

band diagram clearly indicates that there is a wide gap between valence and conduction bands, i.e. forbidden energy gap is very large. It is generally more than 5 eV. Due to the reason of wide gap, it is almost impossible for an electron to cross the gap to go from valence band to conduction band.

At room temperature, the thermal energy is $kT = 20$ mV; therefore, the valence electrons of insulators cannot have so much energy ($kT/2 \ll E_G$) that it is unable to jump to the conduction band from the valence band and hence unable to conduct electric current. It means that insulator has very high resistivity and extremely low conductivity at room temperature. But an insulator may conduct if its temperature is very high or if a very high voltage is applied across it. This situation is known as the breakdown of the insulators.

1.5.2.2 Metals or Conductors

In case of metals (conductors), there is no forbidden energy gap ($E_G = 0$), and the valence and conduction bands overlap. It is shown in Fig. 1.9b. Thus, in a metal, the valence band energies and conduction band energies are the same. An electron in the conduction band experiences almost negligible nuclear attraction. In fact, an electron in the conduction band does not belong to any particular atom and therefore moves freely/randomly throughout the solid. Therefore, a small potential difference across the conductors causes the large number of free electrons to constitute the high electric current. Because of this fact, a metal works as a good conductor.

1.5.2.3 Semiconductors

Figure 1.9c shows the energy band diagram of a semiconductor and it may be observed that forbidden energy gap (E_G) is not very wide. It is 0.7 eV for germanium (Ge) and 1.12 eV for silicon.

At 0 °K (i.e. at an absolute zero temperature of −273 °C), the valence band is usually full and there may be no electron in the conduction band in all elements. It means that no current can flow in the semiconductors also and therefore, the semiconductors behave as the insulators at absolute zero temperature (0 °K). However, in semiconductor materials, both valence and conduction bands are so close (about 1 eV apart) that an electron can be lifted from the valence band to the conduction band by imparting some external energy to it. The external energy applied by the heat at room temperature is not sufficient ($kT = 26$ mV $\ll E_g$) to lift electrons from valence band to conduction band but still some electrons due to this small thermal energy also jump to the conduction band. Hence, at a room temperature, semiconductors are able to conduct some electric current. If temperature is further raised above room temperature, more and more valence electrons acquire energy and cross the energy gap to go to conduction band. Due to this reason, semiconductors have negative temperature coefficient of resistance.

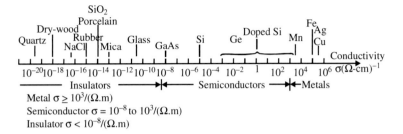

Fig. 1.10 Based on the conductivity value of some materials, they are classified as showing metal, semiconductor and insulator

Some materials are shown in Fig. 1.10 which are classified as metals, semi-conductors and insulators depending upon the value of their conductivity.

A portion of the periodic table in which the most common semiconductors are found is shown in Table 1.1. Silicon and Germanium are in Group IV and are the elemental semiconductors. But the Gallium Arsenide (GaAs) is in a Group III–IV compound semiconductor, and the Silicon Carbide (SiC) is in a Group IV–IV compound semiconductor. In addition to Group III–IV compound semiconductor, Group II–VI compound also forms semiconductors.

The resistivity (inverse of conductivity) of a semiconductor depends on the doping density but is less than an insulator but more than a conductor (Table 1.2).

Table 1.1 A portion of periodic table

S. No.	Group III	Group IV	Group V
1.	B (Boron)	C (Carbon)	
2.	Al (Aluminium)	Si (Silicon)	P (Phosphorus)
3.	Ga (Gallium)	Ge (Germanium)	As (Arsenic)

Table 1.2 Resistivity of some materials for comparison

S. No.	Substances	Nature	Resistivity (Ωm)
1.	Copper	Conductor	1.72×10^{-8}
2.	Germanium	Semiconductor	0.63
3.	Silicon	Semiconductor	2×10^{3}
4.	Glass	Insulator	9×10^{11}
5.	Nichrome	Resistance material	10^{-4}

Table 1.3 The comparison of metals, semiconductors and insulators

S. No.	Parameters of comparison	Metals (Conductors)	Semiconductors	Insulators
1.	Forbidden gap (E_G)	$E_G = 0$ No gap	$E_G < 3$ eV Medium gap	$E_G > 5$ eV Large gap
2.	Effect of temperature on resistance	R-increases as T increases	R-decreases as T increases	R-decreases as T increases
3.	Temperature coefficient of resistance	Positive	Negative	Negative
4.	Conductivity (σ) at room temperature (300 °K)	Very high $\sigma \geq 10^3/(\Omega m)$	Moderate $\sigma = 10^{-8}-10^3/(\Omega m)$	Very low $\sigma < 10^{-8}/(\Omega m)$
5.	Resistivity	Very low	Moderate	Very high
6.	Number of electrons for conduction	Very large ($>10^{23}$ cc)	Moderate ($10^{11}-10^{20}$/cc)	Almost zero or very less
7.	Applications	As conductors, wires, bus bars	Semiconductor devices	Capacitors insulation for wires
8.	Examples	Al, Cu	Si, Ge, GaAs, InP, etc.	paper, mica, glass, etc.

1.6 Comparison of Metals, Semiconductors and Insulators

See Table 1.3.

1.7 Classification of Semiconductor: Intrinsic and Extrinsic

The semiconductor can be classified into two types:

(a) Intrinsic Semiconductor
(b) Extrinsic Semiconductor.

The intrinsic semiconductor refers to pure materials (no impurities or lattice defects) whose electrical conductivity arises due to thermally excited electron into conductivity band and determined by their inherent conductive properties. Pure elemental (e.g. Si and Ge) or compound (e.g. GaAs, GaN, etc.) are intrinsic semiconductivity materials.

The extrinsic semiconductors or impure semiconductors are materials in which a small amount, say part per million (ppm), or a suitable impurity is added to the pure semiconductors, which is usually called doping. This doping is normally taken in number of impurity atoms per cc. This doping changes the properties of semiconductors.

1.8 Intrinsic Semiconductor

A semiconductor is said to be intrinsic only if the impurity content is less than one part impurity in hundred million parts of semiconductors. Si and GaAs are the two widest used intrinsic semiconductors nowadays.

Figure 1.5c shows the two-dimensional crystal structure of Si. There are four electrons in the outermost orbit or valence shell.

Each of the four valence electrons takes part in forming covalent bond (sharing of electrons) with the four neighbouring atoms. A covalent bond consists of two electrons, one from each adjacent atom. Atoms bond together in an attempt to get eight electrons in the valence shell to become a most stable structure (Fig. 1.7c).

1.8.1 Conduction in Intrinsic Semiconductor: Effect of Temperature on Conductivity of Intrinsic Semiconductors

As discussed earlier, any semiconductor (Ge or Si) acts as a perfect insulator at absolute temperature (0 °K). However, at room temperature (300 °K), some electron–hole pairs are produced due to thermal energy. It is seen that in silicon and germanium, the intrinsic carrier concentration (concentration of free electrons or holes) are 1.5×10^{10}/cc and 2.3×10^{13}/cc, respectively, at room temperature (300 °K). It means that the semiconductor has a small conductivity. But if the temperature is raised further, more electron–holes pairs are produced leading to higher concentration of free charge carriers and also the conductivity. Hence, the intrinsic semiconductor has negative temperature coefficient of resistance.

In an intrinsic semiconductor, both the electrons and the holes contribute to the conductivity. Hence, the total conductivity can be written as

$$\sigma = n_i q \left(V_n + V_p \right) \tag{i}$$

As we know that

$$n_i \alpha e^{-E_g/2kT} \tag{ii}$$

where E_g is the energy gap, therefore, we can write

$$\sigma = K_0 e^{-E_g/2kT} \qquad \text{(iii)}$$

where K_0 is some constant. This shows that as the temperature increases, the conductivity of the semiconductors increases exponentially. This is because the density of electron–hole pairs increases.

If σ_∞ be the (extrapolated) conductivity for $T = \infty$, then

$$K_0 = \sigma_\infty \qquad \text{(iv)}$$

Therefore, equation of σ becomes

$$\sigma = \sigma_\infty e^{-E_g/2kT} \qquad \text{(v)}$$

Taking natural $\log_e$ on the both sides, we get

$$\log_e \sigma = \log_e \sigma_\infty - \frac{E_g}{2kT} \qquad \text{(vi)}$$

A plot between $\log_e \sigma$ and $1/T$ gives a straight line whose slope is $\left(\frac{-E_g}{2k}\right)$ as shown in Fig. 1.11.

1.8.2 Formation of Free Electrons and Holes as Free Carriers

The breaking of covalent bond (a) is equivalent to raising an electron from valence band to the conduction band; (b) is equivalent to moving out of electron from the atom, making it a free electron; and (c) leads to the formation of a hole (i.e. positive charge or deficiency). The energy required to break a covalent bond is same as the forbidden energy E_G and equals 0.72 eV for Ge and 1.12 eV for Si. This energy

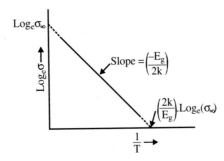

Fig. 1.11 Plot between $\log_e \sigma$ and $1/T$

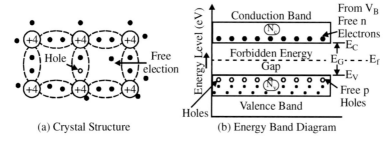

Fig. 1.12 Formation of Electron–Hole Pair in an Intrinsic Semiconductor. Probability $f(E)$ of an electron to occupy Fermi energy level E_f is ½, i.e. $f(E_f)$ = ½ (See next Sect. 1.8.3)

may be supplied either (i) by raising the temperature, thereby imparting thermal energy, (ii) by incidence of photons in the visible range or (iii) by bombardment of the semiconductor materials by the α-particles, electrons and X-rays. Thus, radiation of photons or electrons having energy > Eg, free electrons (represented by solid circle) and holes (represented by a small circle) are always generated in pairs. So concentration of free electrons and holes will always be equal in intrinsic semiconductor. Such thermal generation of free electron–hole pairs using crystal structure and energy band diagram is shown in Fig. 1.12a and b, respectively.

Since a crystal is electrically neutral, whenever an electron moves out of an atom, a hole is created behind with a net possible charge equal in magnitude to that of negative charge of free electrons. The concept of hole as a positively charged particle merely helps in simplifying the explanation of conduction in semiconductors.

1.8.3   Carrier Concentration in Intrinsic Semiconductors

Here, we use the concept that electrons in solids follow Fermi–Dirac statistics, i.e. the probability of an electron to occupy an available state at an energy level (E) at thermal equilibrium at temperature $(T\ °K)$ is given by

$$f(E) = 1/\left[1 + \exp\{(E - E_f)/kT\right\}\tag{1.1}$$

where

k Boltzman's constant.
E_f Fermi's energy level of electron.

Thus, probability of an electron having energy $E = E_f$ is half, i.e.

$$f(E) = \frac{1}{1+1} = 0.5 \tag{1.1a}$$

and this becomes the definition of Fermi level. It is exactly between E_c and E_V, is intrinsic semiconductor (see Fig. 1.12).

Using the above probability distribution function $F(E)$, we can prove that the concentration of electron (n) and holes (p) in semiconductor will be (Fig. 1.12)

$$n = N_c \cdot \exp\left[-\left(E_c - E_f\right)/kT\right] \tag{1.2a}$$

$$p = N_v \cdot \exp\left[-\left(E_f - E_v\right)/kT\right] \tag{1.2b}$$

where

E_f = Fermi energy level in semiconductors.
$N_c = 2(2\pi kTm_e/h^2)^{3/2}$
 = Effective density of energy states available in the conduction band for electrons = 2.8×10^{19}/cc.
$N_v = 2(2\pi kTm_h/h^2)^{3/2}$
 = Effective density of available energy states in valence bands for holes
 = 1.02×10^{19}/cc (different as $m_e \neq m_h$).
m_e, m_h = Effective mass of electrons and holes.
N_D, N_A = Donor and acceptor doping concentrations.

Since electron and hole concentrations are equal in an intrinsic semiconductor $n = p = n_i = p_i$; therefore, Eq. (1.2a) and (1.2b) leads to (with $E_f = E_{fi}$)

$$n_i = N_c \cdot \exp\left[-\left(E_c - E_{fi}\right)/kT\right] \tag{1.3}$$

$$p_i = N_v \cdot \exp\left[-\left(E_{fi} - E_v\right)/kT\right] \tag{1.4}$$

and with $n_i = p_i$ in Eq. (1.3), we get the intrinsic Fermi level energy E_{fi} as

$$E_f = E_{fi} = \frac{E_c + E_v}{2} - \frac{kT}{2}\ln\left(\frac{N_c}{N_v}\right) \tag{1.5}$$

Thus at 0 °K, $E_{fi} = \frac{E_c + E_v}{2}$, i.e. midpoint of band gap with E_C, E_v, $E_f \approx -10$ to -11 eV, in energy band diagram Fig. 1.6 of silicon. At room temperature ($T = 300$ °K), as $N_c \neq N_v$, the second term of Eq. (1.5), i.e. $\frac{kT}{2}\ln\left(\frac{N_c}{N_v}\right)$, is rather small (a few meV) which is neglected, and therefore here also E_{fi} is assumed to be located approximately in the middle of the band gap for intrinsic semiconductor. Also, by the product ($n_i \cdot p_i$) from Eq. (1.3) and Eq. (1.4), we get

Fig. 1.13 Intrinsic carrier densities (n_i) in Ge, Si and GaAs as a function of temperature (T). At room temperature ($T = 300\,°K$), n_i and E_g are also shown

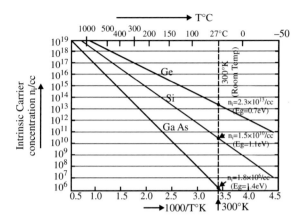

$$n_i^2 = n_i \cdot p_i = N_c N_v \cdot \exp[-(E_c - E_v)/kT] \tag{1.6}$$

$$n_i = \sqrt{N_c N_v}\, \exp\left(-E_g/2kT\right) \tag{1.7}$$

where

$E_g = (E_c - E_v)$ Energy gap.

On putting the above values of N_c and N_v in Eq. (1.7), we see that

$$n_i = 2\left(2\pi kT/h^2\right)^{3/2} (m_e m_h)^{3/4} \cdot \exp\left(-E_g/2kT\right) \tag{1.8}$$

$$n_i = K_1 \cdot T^{3/2} \exp\left(-E_g/2kT\right) \tag{1.9}$$

where

$$K_1 = 2\left(2\pi k/h^2\right)^{3/2} \cdot (m_e m_h)^{3/4} = \text{constant}$$

Thus n_i is a strong variable of E_g and T. It increases with temperature (T) but reduces with band gap as depicted in Fig. 1.13 computed from the above equation.

1.9   Extrinsic Semiconductors: Donor and Acceptor Levels

As mentioned earlier in Sect. 1.7 that at room temperature, intrinsic semiconductor (Silicon) has only about 10^{10} electrons per cubic centimetres which contribute to the conduction of the electric current, and hence these are of no practical use, being close to insulators. If a small amount of a pentavalent or trivalent impurity is introduced into a pure Germanium (or Silicon) crystal, then the conductivity of the crystal increases appreciably and the crystal becomes an 'extrinsic semiconductor'.

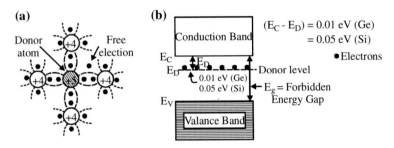

Fig. 1.14 **a** Effect of doping pure Ge (or Si) crystal with a pentavalent impurity. **b** Donor level in *n*-type semiconductor

The process of adding small amounts of impurities to the semiconductor material is called doping. The impurity elements that replace a regular lattice atom are known as dopants.

When a small amount of a **pentavalent** (Antimony, phosphorus and arsenic) atom (e.g. 0.0001%) is added, then (a) four of its valence electrons form regular electron pair bonds with their neighbouring Ge (or Si) atoms as shown in Fig. 1.14a, b, while the fifth electron gets loosely bonded to silicon with a donor level the binding energy E_D (or lattice energy) of about 0.01 eV for Ge and 0.05 eV for Si just below the conduction band. At room temperature with thermal energy of $kT = 0.026$ eV being greater than or close to E_D (Fig. 1.14b), this donor level electron gets dissociated from it and moves through the crystal as a conduction electron and is therefore called 'donors'. Therefore, the crystal becomes '*n*-type Semiconductors' because the conduction in this is predominately by negative charge carriers (free electrons).

At room temperature, the extra electrons (fifth electrons) of almost all the donor, atoms are thermally activated from the donor level into the conduction band, where they move as charge carriers, when a voltage is applied to the crystal, leaving the E_D level vacants.

When a **trivalent** (Boron, aluminium, gallium or indium) atom replaces a Ge (or Si) atom in a crystal lattice as shown in Fig. 1.15a, only three valence electrons are

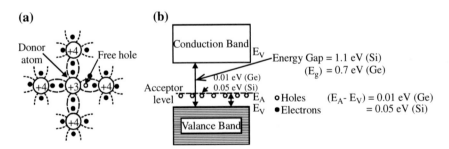

Fig. 1.15 **a** Effect of doping pure Ge (or Si) crystal with or trivalent impurity. **b** Acceptor level in extrinsic *p*-type semiconductor at 0 °K

available to form covalent bands with the neighbouring Ge (or Si) atoms. This results in an empty space or a '*hole*'. When a voltage is applied to (a) crystal, then an electron bound to a neighbouring Ge (or Si) atoms moves out and occupies the existing hole position, therefore creating a new hole. Thus, the conduction mechanism in these semiconductors with acceptor impurities is predominated by positive carriers which are introduced into valence band. This type of semiconductor is called '*p*-type semiconductor'.

The trivalent impurity atoms introduce vacant discrete energy levels (acceptor level E_A) just above the top of valence band as shown in Fig. 1.15b. These are called 'acceptor impurity levels' which are only 0.01 eV above the valence band for Ge and 0.05 eV for Si at 0 °K. At room temperature (just like donors), almost all the holes move to the valence band vacating the E_A level.

1.9.1   Space Charge Neutrality and n-Type/p-Type *Semiconductor*

The mass action law $(n \cdot p = n_i^2)$ gives the relationship between free electron and hole concentrations as will be proved in Sect. 1.11. These concentrations are further related by the law of electrical neutrality as follows.

Let

N_D Concentration of donor atoms or number of positive charges/m^3 contributed by donor ions after the atom loses one electron.

p Hole concentration.

p_p, p_n Majority and minority hole concentrations in *p*-type and *n*-type semiconductors.

N_A Concentration of acceptor atoms, i.e. –ve charge ions.

n Free electron concentration.

n_n, n_p Majority and minority electron concentrations in *n*-type and *p*-type semiconductors.

For any semiconductor, total positive charge density or total number of positive charges/m^3 is $(N_D + p)$.

Similarly, total negative charge density or the total number of negative charges/m^3 is $(N_A + n)$.

According to the principle of electrical neutrality, we know that

$$N_D + p = N_A + n \qquad (1.10)$$

For *n*-type semiconductor,

$$N_A = 0 \ (\because \text{concentration of acceptor ion is zero in } n\text{-type})$$

and

$$n \gg p \; (\because \text{free electron are much larger in number than holes})$$

Using the above two assumptions in Eq. (1.10), we get

$$N_D = n \tag{1.11}$$

Hence, for an n-type semiconductor, the concentration of free electrons is equal to the concentration of donor ions.

To avoid any confusion, let us write n_N in place of n concentration for free electrons, which is n-type semiconductor; so Eq. (1.11) is modified into

$$N_D = n_N = \text{electron majority in } n\text{-type} \tag{1.12}$$

For n-type semiconductor, mass action law $n \cdot p = n_i^2$ can be rewritten (discussed in Sect. 1.11.2) for obtaining the minority hole concentration:

$$n_N \cdot p_N = n_i^2 \tag{1.13}$$

So from Eq. (1.12)

$$p_N = \frac{n_i^2}{n_N} = \frac{n_i^2}{N_D} = \text{(holes minority in } n\text{-type)} \tag{1.14}$$

Similarly, for p-type semiconductors

$$N_D = 0 \; (\text{concentration of donor ion is zero in } p\text{-type.})$$

and

$$p \gg n$$

So Eq. (1.10) becomes

$$p = N_A = \text{hole majority in } p\text{-type} \tag{1.15}$$

As like n-type semiconductor, to avoid any confusion, let us also write p_p in place of p concentration for holes in p-type semiconductor.

So,

$$p_P = N_A = \text{majority holes} \tag{1.16}$$

For p-type semiconductor, mass action law can be written as

$$n_P \cdot p_P = n_i^2 \tag{1.17}$$

Putting the value of p_P of Eq. (1.16) in Eq. (1.17), we get the electron minority carriers in p-semiconductor

$$n_P = \frac{n_i^2}{N_A} = \text{Concentration of free electrons minority in } p\text{-type semiconductor}$$

(1.18)

1.10 Fermi Level

As discussed in Sect. 1.8.3, the Fermi level is defined as the energy state or level consisting of 50% probability of being filled by an electron (see in Eq. 1.1a) (Fig. 1.16).

For intrinsic semiconductor, the Fermi level is independent of temperature and it lies in the middle of the forbidden energy gap. In other words, Fermi level is at the centre of two bands, i.e. valence band and conduction band.

In n-type semiconductor, the free electrons are majority carriers, whereas the holes are the minority carriers as shown in Fig. 1.17a. Similarly, for p-type semiconductor, the holes are majority carriers, whereas the free electrons are the minority carriers, as shown in Fig. 1.17b.

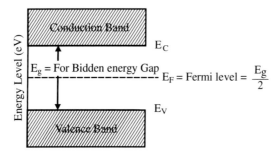

Fig. 1.16 Fermi level in an intrinsic semiconductor

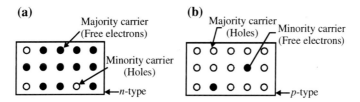

Fig. 1.17 a Extrinsic n-type (Donor) semiconductor. b Extrinsic p-type (Acceptor) semiconductor

For an n-type doping in intrinsic semiconductor, the number of electrons increases in the conduction band because of donor impurities. Due to this reason, the Fermi level shifts upward from the centre closer to the conduction band as shown in Fig. 1.7a and c at 0 °K and 300 °K, respectively. Fermi level for n-type semiconductor material is also temperature-dependent and expressed at very low temperature as

$$E_{FN} = E_{fn} = \left(\frac{E_C + E_D}{2}\right) - \frac{kT}{2}\log_e\left(\frac{2n_C}{N_D}\right) \tag{1.19}$$

For a p-type doping in intrinsic semiconductor, the number of electrons decreases in the conduction band because of acceptor impurities. Due to this reason, the Fermi level is lowered from the centre and becomes closer to the valence band as shown in Fig. 1.18b and d for 0 °K and 300 °K, respectively.

For a p-type semiconductor material, the Fermi level is also temperature-dependent as

$$E_{fp} = E_{FP} = \left(\frac{E_V + E_A}{2}\right) + \frac{kT}{2}\log_e\left(\frac{2n_v}{N_A}\right) \tag{1.20}$$

where

E_C	Fermi level of intrinsic semiconductor.
$E_{FP} = E_{fp}$	Fermi level for p-type.
$E_{FN} = E_{fn}$	Fermi level for n-type.
k	Boltzmann's constant $= 1.38 \times 10^{-23}$ J/°K.
T	Temperature in °K.
E_V	Energy level in the valence band.
E_C	Energy level in the conduction band.
E_D	Donor energy level.
E_A	Acceptor energy level.
$N_C = n_C$	Number of electron concentrations in conduction band.
$N_V = n_V$	Number of electron concentrations in valence band.
N_D	Concentration of donor impurity.
N_A	Concentration of acceptor impurity.

At 0 °K, Eqs. (1.20) and (1.21a, 1.21b) become

$$E_{fn} = E_{FN} = (E_C + E_D)/2 \tag{1.21a}$$

$$E_{fp} = E_{FP} = (E_V + E_A)/2 \tag{1.21b}$$

i.e. at 0 °K, the Fermi level for n-type semiconductor is at the midpoint of conduction band edge and donor energy level, while for p-type at the midpoint of valence band edge and acceptor level.

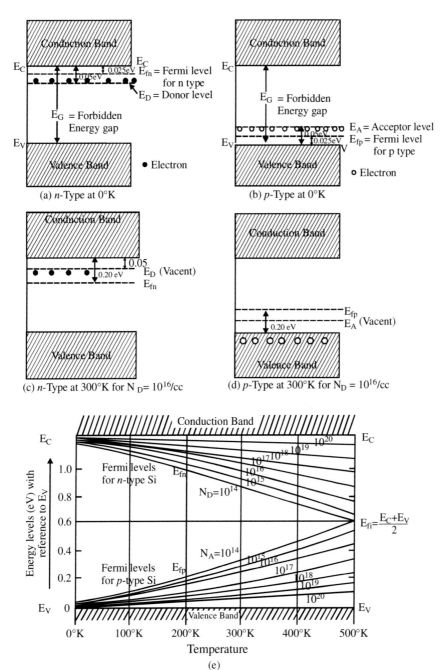

Fig. 1.18 Fermi levels in extrinsic semiconductor (Si). **a** *n*-type at 0 °K, **b** *p*-type at 0 °K, **c** *n*-type at 300 °K, **d** *p*-type at 300 °K, **e** E_{pn} and E_{fp} as a function of temperature and impurity doping

At higher temperature (say at room temperature 300 °K), all the donor electrons jump to the conduction band and all the acceptor holes to the valence band, making free carriers. Here, all the donor atoms get ionized to free its electron. Therefore, carrier densities $n = N_D$ and $p = N_A$. This way, the donor and acceptor energy levels get vacated (see Fig. 1.18c, d).

The Fermi levels of electrons and holes in n-type and p-type semiconductors (f_n, f_p) can have two types of expression:

(a) From Eqs. (1.19) and (1.20), on putting the value with

$$n = N_D \text{ at } E_f = E_{fn}$$

and $$p = N_A \text{ at } E_f = E_{fp} \qquad \text{as}$$

$$E_{FN} = E_{fn} = E_C - \frac{kT}{2} \log_e \left(\frac{2N_C}{N_D} \right) \qquad (1.21c)$$

$$E_{FP} = E_{fp} = E_V - \frac{kT}{2} \log_e \left(\frac{2N_V}{N_A} \right) \qquad (1.21d)$$

(b) On dividing the electron carrier densities' equations, Eq. (1.19) for intrinsic (n_i) and Eq. (1.3) for a general extrinsic semiconductor (n), we get by putting $E_f = E_{fn}$

$$\frac{n}{n_i} = \frac{N_C \cdot \exp\left[-(E_C - E_{fn})/kT\right]}{N_C \cdot \exp\left[-(E_C - E_{fi})/kT\right]}$$
$$\frac{n}{n_i} = \exp\left[(E_{fn} - E_{fi})/kT\right] \qquad (1.22)$$

Similarly, we can show that for hole carrier densities as

$$\frac{p}{p_i} = \exp\left[(E_{fp} - E_{fi})/kT\right] \qquad (1.23)$$

Equations (1.22) and (1.23) give the Fermi energy definition for n-type and p-type semiconductor as

$$E_{fn} = E_{FN} = E_{fi} + kT \log_e \left(\frac{n}{n_i} \right) = E_{fi} + kT \log_e \left(\frac{N_D}{n_i} \right) \qquad (1.24)$$

$$E_{fp} = E_{FP} = E_{fi} - kT \log_e \left(\frac{p}{p_i} \right) = E_{fi} - kT \log_e \left(\frac{N_A}{p_i} \right) \qquad (1.25)$$

We also see from Eqs. (1.19) and (1.20) that the Fermi levels E_{fn} and E_{fp} move:

(a) Away from their respective bands at higher temperatures closer to the midpoint E_{fi} (as shown in Fig. 1.18e).
(b) Towards their respective bands with higher doping densities, increasing the carriers.

Finally, when the doping level in Si becomes greater than 10^{19}/cc, then the Fermi level merges with the conduction band for n-type and with the valence band for p-type semiconductor. Such semiconductor starts behaving like a conductor in reverse bias losing the rectifying properties of pn junction. But in the forward bias between 0.05 and 0.5 V or so, it has −ve resistance and therefore these diodes are called tunnel diodes and are used as microwave oscillators.

Thus we see from Fig. 1.18 that the gaps of Fermi levels from their respective bands, i.e. $(E_{fn} - E_c)$ for *n*-type semiconductors and $(E_{fp} - E_v)$ for *p*-type semiconductors are as follows:

$$\left|E_{fn} - E_c\right| = \left|E_{fp} - E_v\right| = 0.05/2 \, \text{eV at} \, 0\,°\text{K for} \, N_D \, \text{or} \, N_A = 10^{15}/\text{cc}$$
$$= 0.25 \, \text{eV at} \, 300\,°\text{K for} \, N_D \, \text{or} \, N_A = 10^{15}/\text{cc}$$
$$= 0.20 \, \text{eV at} \, 300\,°\text{K for} \, N_D \, \text{or} \, N_A = 10^{16}/\text{cc}$$
$$= 0.02 \, \text{eV at} \, 300\,°\text{K for} \, N_D \, \text{or} \, N_A = 10^{19}/\text{cc}$$

When a semiconductor has equal doping of *n*-type and *p*-type (i.e. $N_A = N_D$), then the Fermi level is at the E_i level and the semiconductor is said to be compensated semiconductor; such devices can be made by ion implantation technology only, where original doping distribution does not get disturbed.

1.11   Drift and Diffusion Currents

The current through a semiconductor material is of two types: Drift and Diffusion.

(a) When an electric field is applied across the semiconductor, the charge carriers attain a drift velocity v_d. This drift velocity v_d is equal to the product of the mobility of charge carriers and the applied electric field intensity E is created by the *dc* supply/battery as shown in Fig. 1.19. The holes drift / move towards the negative terminal of the battery and electron towards the positive terminal. This combined effect of movement of the charge carriers constitutes a current known as **drift current**.

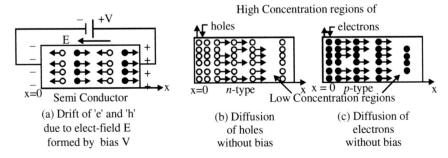

Fig. 1.19 a Drift of electrons and holes in semiconductors. **b, c** Diffusion is due non-zero concentration gradient (dN/dx) even within one type (*p* or *n*) of semiconductor

So, the equation for the drift current density J_n due to free electrons may be given as

$$J_n = q n \mu_n E \tag{1.26}$$

Similarly, the equation for the drift current density J_p due to holes concentration is given by

$$J_p = q p \mu_p E \tag{1.27}$$

Here, it may be noted that for *n*-type semiconductors $n \gg p$, therefore $J_n \gg J_p$ and is the other way round for *p*-types, where

q electronic charge.
μ_n Mobility of free electrons.
μ_p Mobility of holes.
n Free electrons concentration.
p Holes concentration.
E Applied electric field.

(b) It is observed that charge carriers flows in a semiconductor even in the absence of an applied electric field, if a concentration gradient exists when the number of either free electrons or holes is greater in one region of a semiconductor as compared to the rest of the region. It is found that if the concentration gradient of charge carriers exists in a semiconductor, the carriers tend to move from the region of higher concentration to the region of lower concentration. This process is called diffusion, and the electric current due to this process is called **diffusion current**.

So, the total current density in n-type or p-type concentration is equal to sum of drift current density and diffusion current density.

$$\text{Total current} = \text{Drift current} + \text{Diffusion current} \tag{1.28}$$

1.11.1   Conventional Current and Electron Flow

(i) **Conventional current**: The current flow from positive to negative is referred to as conventional current. Every graphical symbol used to represent an electronic device has an arrowhead which indicates conventional current direction. Therefore, the electronic circuits are most easily explained by using the conventional current flow.

(ii) **Electron flow**: Under the influence of the external dc source or battery, the free electrons in the semiconductor slab, which are negatively charged, will get attracted towards the positive terminal, and holes being positively charged will be attracted towards the negative terminal of the external battery. Thus, as shown in Fig. 1.20a, the electrons flow exactly in the opposite direction to that of the conventional current. The study of electron movement is essential in understanding the operation of any electronic device.

1.11.2   Drift Current, Mobility, Conductivity and Law of Mass Action

The conductivity of a material is proportional to the concentration of free electrons. When we apply a constant electric field E to a metal or a semiconductor, the free electrons would be accelerated and the velocity would increase. However, because

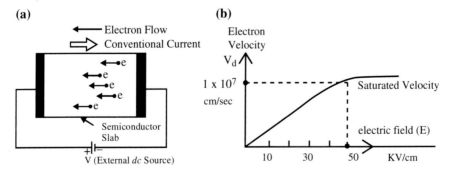

Fig. 1.20   a Conventional current and electron flow in semiconductor. **b** Electron velocity saturation with electric field in Si

of the collision of electrons, electrons lose energy and a steady-state condition is reached, where a finite value of drift velocity v_d is attained. The drift velocity v_d is in a direction opposite to that of the electric field, and its magnitude is proportional to E up to a saturation limit of 1×10^7 cm/s which is reached at $E = 50$ kV/cm for silicon (Fig. 1.20b).

Hence,

$$v_d = \mu E \tag{1.29}$$

where $\mu =$ mobility of electrons in m²/volt-second.

Thus mobility may be defined as the average particle drift velocity per unit electric field.

Therefore, due to the applied electric field, a steady-state drift velocity has been superimposed upon the random thermal motion of the electrons. If the concentration of free electrons is n (electrons per cubic metre), the current density J in amperes per square metre is

$$
\begin{aligned}
J &= nqv_d \\
J &= nq\mu E \quad \text{(using } v_d = \mu E) \\
\text{or} \quad J &= \sigma E \\
\text{where} \quad \sigma &= nq\mu
\end{aligned}
\tag{1.30}
$$

Here, σ is the conductivity of the metal in (ohm-metre)$^{-1}$. For a good conductor, n is very large. It is approximately 10^{22} electrons/cm³. Equation (1.30) may be referred to as Ohm's law which states that the conduction current density is proportional to the applied electric field.

Intrinsic semiconductors behave as perfect insulator at absolute zero (0 K) because at 0 K, the valence band remains full, the conduction band empty and no free charge carriers are available for conduction as shown in Fig. 1.18. But at room temperature (300 K), the thermal energy is sufficient to create a large number of electron–hole pairs. If an electric field is applied, the current flows through the semiconductor. The current flows in a semiconductor due to the movement of electrons in one direction and holes in the opposite direction. These two current densities are given by

$$J_n = q \cdot n \cdot \mu_n \cdot E \text{ and } J_P = q \cdot p \cdot \mu_p \cdot E \tag{1.31}$$

where

q	Charge on an electron and hole being equal.
n, p	Electron and hole concentrations in intrinsic semiconductor, respectively.
μ_n, μ_p	Mobilities of electrons and holes in semiconductor, respectively.
E	Electric field applied.

The total current density will be the sum of above two drift currents, i.e.

$$J = J_n + J_p \tag{1.32}$$

$$J = qn\,\mu_n E + qp\,\mu_p E$$
$$J = q\,E\left(n\mu_n + p\,\mu_p\right) \tag{1.33}$$

$$J = \sigma E \tag{1.34}$$

where

$$\sigma = \left(n\mu_n + p\mu_p\right)q$$

Here, σ is called the conductivity of semiconductor in mhos ($\mho$). But for intrinsic semiconductor, we know that the number of electrons is equal to number of holes called intrinsic carrier concentration (n_i), i.e.

$$n = p = n_i$$

Putting $n = p = n_i$ in Eq. (1.33), we get

$$J = qE\left(n_i\mu_n + n_i\mu_p\right) \tag{1.35}$$

Therefore, the conductivity of intrinsic semiconductor will be J/E as

$$\sigma_i = q\left(n_i\mu_n + n_i\mu_p\right)$$
$$\sigma_i = qn_i\left(\mu_n + \mu_p\right) \tag{1.36}$$

Hence, the conductivity of an intrinsic semiconductor depends upon its intrinsic concentration (n_i), mobility of electron (μ_n) and mobility of holes (μ_p).

But for n-type semiconductor, we know that $n \gg p$ and $p \gg n$ for p-type. Therefore, the conductivity of n-type and p-type semiconductor are, respectively,

$$\sigma_n = q \cdot n \cdot \mu_n \text{ and } \sigma_p = q \cdot p \cdot \mu_p \tag{1.37}$$

When a pure semiconductor is doped with n-type impurities, the number of electrons in the conduction band increases above a level and the number of holes in the valence band decreases below a level which would have been available in the intrinsic or pure semiconductor. Similarly, if p-type impurities are added to a pure semiconductor, the number of holes increases in the valence band above a level and the number of electrons decreases below a level which would have been available in the intrinsic semiconductor.

Under thermal equilibrium for any semiconductor, the product of the number of holes and the number of electrons is constant and is independent of the amount of donor and acceptor impurity doping. This relation is known as mass action law as shown in Eqs. (1.18) and (1.19).

Mathematically, we have written as

$$n \cdot p = n_i^2 \tag{1.38}$$

where

n number of free electrons per unit volume.
p number of holes per unit volume.
n_i the intrinsic carrier concentration.

Mobility (μ) varies as T^{-m} over a temperature range of 100–400 °K. For silicon (Si), $m = 2.5$ for electrons and 2.7 for holes. For germanium (Ge), m = 1.66 for electrons and 2.33 for holes. The mobility is also found to be a function of electric field intensity and remains constant only if $E < 10^3$ V/cm in n-type silicon. For $10^3 < E < 10^4$ V/cm, the mobility (μ_N) varies approximately as $E^{-1/2}$. For higher fields in silicon, the mobility (μ_N) is inversely proportional to E and the carrier speed approaches a constant value of 10^7 cm/s.

1.11.3 Diffusion Current, Diffusion Length and Einstein Relation

As discussed in Sect. 1.11, the diffusion of carriers is caused by the difference in the carrier density in one region than the other in a semiconductor (Fig. 1.19). This diffusion is similar to the case of a drop of dye in water spreads out from (Refer Sect. 1.12) point of higher concentration to lower one or a perfume bottle, when opened in a room spreads all over. This variation of concentration [i.e. concentration gradient (dN/dx)] is due to any of the following reasons:

(a) Doping of impurity is not uniform.
(b) Temperature of the semiconductor is not uniform along the length (x).
(c) Some radiation (like X-ray, etc.) is falling at one end and not on the other end.

In case (a), the charge carriers will leave the impurity ionized and diffuse to the lower doping region, where the carriers are less. This process reaches a steady state very soon. Along with dN/dx being non-zero, the Fermi level also changes with x due to impurity gradient $\left(\mathrm{d}E_p/\mathrm{d}x \neq 0\right)$ (see Fig. 1.19b, c).

In case (b), higher temperature region has larger population of carriers and diffusion current reaches a steady state so far as difference of temperature is constant, with the two ends shorted by a wire. In case (c), also steady state reaches so far as the radiation intensity is constant. In all the above case, once the carrier distribution becomes uniform, diffusion stops. If we are able to maintain the difference in carrier density (i.e. d$N/\mathrm{d}x \neq 0$), by the way of extracting it from the other, diffusion and diffusion current (J_d) will continue. This is possible by connecting the two ends with an external wire.

Due to this movement of carriers from one region to another because of diffusion, it is called as diffusion current (J_d), which is proportional to the concentration gradient, i.e.

$$J_{dn} \propto dn/dx|\ldots\text{For electrons diffusion}$$
$$J_{dp} \propto dp/dx\ldots\text{For hole diffusion}$$

Therefore,

$$J_{dn} = q \cdot D_n(dn/dx)$$

and

$$J_{dp} = +q \cdot D_p(dp/dx)$$

where D_n and D_p are diffusion constants in cm^2/s.

Thus, the total current (electrons and holes) being due to drift and diffusion, can be expressed as

$$J_n = n \cdot q \cdot \mu_n \cdot E + q \cdot D_n \cdot \left(\frac{dn}{dx}\right) \tag{1.39}$$

$$J_p = p \cdot q \cdot \mu_p \cdot E + q \cdot D_p \cdot \left(\frac{dp}{dx}\right) \tag{1.40}$$

Einstein's relation: For a non-uniformly n-type doped semiconductor $n = n(x)$, we use the equation derived in Sect. 1.10 for $n(x)$ in terms of n_i, E_{fn}, E_{fi} as

$$n(x) = n_i \exp\left[(E_{fn} - E_{fi})/kT\right]$$
$$E_{fn} - E_{fi} = kT \ln\left[\frac{n(x)}{n_i}\right] \tag{1.41}$$

We also know that in such a sample, distance of E_{fn} from the conduction band will change with distance along the length but will remain constant along the length of the sample, while E_{fi} will remain in the middle (see Fig. 1.21c) of the band gap. Therefore, as $dE_{fn}/dx = 0$, the differential of above equation gives

$$-\frac{dE_{fi}}{dx} = \left\{\frac{kT}{n(x)} \cdot \frac{dn(x)}{dx}\right\} \tag{1.42}$$

Also potential (ϕ) across the sample will be

$$\phi = \frac{1}{q}\left[E_{fn} - E_{fi}\right] \tag{1.43}$$

Fig. 1.21 a, b Variation of carrier concentration along the length of semiconductor, **c** Band diagram where electron energy levels E_c, E_v for *n*-type increase with x along the length

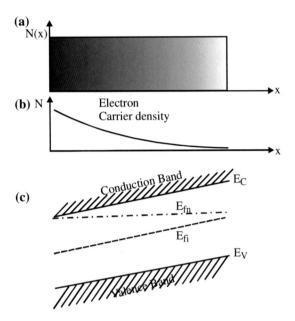

We get electric field by differentiating equation (1.43)

$$\frac{d\phi}{dx} = -E(x) = -\frac{1}{q}\frac{dE_{fi}}{dx}$$

$$E(x) = \frac{1}{q}\left(\frac{dE_{fi}}{dx}\right) \tag{1.44}$$

On putting $\frac{dE_{fi}}{dx}$ of Eq. (1.42) in Eq. (1.44), we get

$$E(x) = \frac{-1}{q}\frac{kT}{n(x)} \cdot \left[\frac{dn(x)}{dx}\right] \tag{1.45}$$

Thus, non-uniform doping leads to an electric field across it. Now at equilibrium, net $J_n = 0$, therefore

$$0 = n \cdot q \cdot \mu_n E(n) + q D_n \left[\frac{dn}{dx}\right] \tag{1.46}$$

On putting the value of $E(x)$ of Eq. (1.45) in Eq. (1.46), we get

$$0 = n(x) \cdot q \cdot \mu_n \cdot \left[\frac{-1}{q} \frac{kT}{n(x)} \cdot \frac{dn(x)}{dx} \right] + q \cdot D_n \cdot \frac{dn(x)}{dx}$$

$$0 = \left[\frac{-\mu_n kT}{q} + D_n \right] q \frac{dn(x)}{dx}$$

$$0 = \frac{-\mu_n kT}{q} + D_n \qquad\qquad (1.47)$$

$$D_n = \frac{\mu_n kT}{q}$$

$$\therefore \quad \frac{D_n}{\mu_n} = \frac{kT}{q}$$

Similarly

$$\frac{D_p}{\mu_p} = \frac{kT}{q} \qquad\qquad (1.48)$$

Thus

$$\frac{D_n}{\mu_n} = \frac{D_p}{\mu_p} = \frac{kT}{q} = \text{Constant at temperature } (T)$$

Above equation is known as Einstein relation. Here, $\frac{kT}{q} = V_T$ is equal to the voltage equivalent of temperature ($T = 300\ °K$) i.e 26 mV.

Diffusion Length: It is defined as the distance travelled (L_n) by a free carrier before recombination in time τ_n. It can be proved for electrons and holes as

$$L_n = \sqrt{\tau_n D_n} \text{ and } L_p = \sqrt{\tau_p D_p}$$

1.11.4   Using Diffusion Concept for Identifying n- or p-Type Semiconductor

As we know that at higher temperature, the number of carriers increases, more and more electrons jump to the conduction band. As a result, resistivity (ρ) and resistance (R) fall with rise of temperature, and hence conductivity (σ) increases. Therefore, $\frac{d\rho}{dT} = $ Negative.

Now if we keep the temperature of one end of a p-type semiconductor high, then at that end excess holes will be there, which will diffuse to the colder end where carriers are less. If this colder end is connected to the hot end by an external wire through a micro-ammeter (μA), then the micro-ammeter (μA) will show the current deflection. However, this current is very small (in μA or less) as it is shown in Fig. 1.22a.

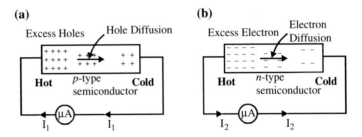

Fig. 1.22 **a** Diffusion of holes from hot region to cold region in *p*-type, **b** diffusion of electrons from hot region to cold region in *n*-type. Conventional current in this *n*-type semiconductor is in the opposite direction as compared to *p*-type semiconductor

Similarly, for an *n*-type semiconductor, conventional diffusion current will be there in the opposite direction. This gives the indication of whether the sample is of *p*-type of *n*-type semiconductor.

1.11.5 High Field and Electron Velocity Saturation

Drift velocity of carriers in semiconductors is proportional to the applied electric field (E = Voltage/d), up to a limit of E_{sat} with the proportionality constant as mobility μ ($v = \mu E$). This limit is E_{sat}, the electric field after which the additional field energy is transferred to the lattice rather than in increasing the velocity. Therefore, velocity approaches a saturation velocity (v_{sat}), also called as the scattering limit velocity. For Si, $E_{sat} \approx 50 \times 10^3$ V/cm and $v_{sat} \approx 10^7$ cm/s. Figure 1.23 shows the plot for electron velocity as a function of electric field.

Fig. 1.23 Electron velocity as a function of electric field in Si

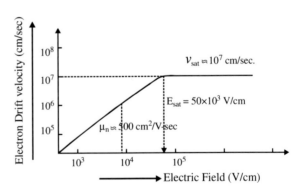

1.12 Hall Effect and Other Properties of Semiconductors

1.12.1 Hall Effect to Measure Magnetic Field Strength or to Identity Between n/p-Type Semiconductors

This property is used for (a) identifying whether a semiconductor is *p*-type or *n*-type, (b) measuring carrier concentration, (c) measuring mobility of charges and (d) measuring magnetic field strength if (a), (b) and (c) are known.

If a current (I_x) in semiconductor is along *x*, direction as in Fig. 1.24 and is placed in a magnetic field (B_z), then a voltage is generated in the third direction (y), perpendicular to both. This voltage is called Hall voltage (V_{Hy}). This is because of the Lorentz force $[F_y = q\ v_x\ B_z]$ being exerted on the charge (q) carried by the current (I_x) with velocity (v_x). The direction of this force (F_y) is along negative *y*-direction for positive charges, i.e. for *p*-type semiconductor. This direction of force is same as per the Fleming's left-hand rule for a current-carrying conductor.

The force F_y in –ve *y*-direction for positive charges will make them accumulate on face *A* for *p*-type, making it positively charged. This charge leads to voltage across *AB* (as V_{Hy}) and consequently electric field (E_y) across the face A_B. This electric field exactly balances the Lorentz force, therefore

$$F_y = qE_y = qv_xB_z$$
$$\therefore\quad E_y = v_xB_z \tag{1.49}$$

This E_y is called as Hall field and corresponding voltage across *AB* (V_{Hy}) as hall voltage.

$$V_{Hy} = E_y \cdot W = (v_xB_z)W$$

$$\therefore\quad v_x = \frac{V_{Hy}}{(B_z \cdot W)} \tag{1.50}$$

Fig. 1.24 Hall effect voltage direction for identifying *p*-type or *n*-type

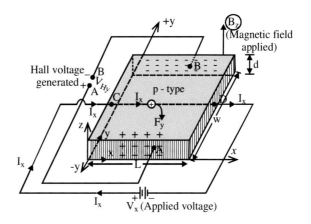

Thus for a *p*-type semiconductor, V_{Hy} will be +ve while for *n*-type semiconductor, it will be −ve and thus it can indicate whether a semiconductor is *p*-type or *n*-type.

Now the current (I_x) can be written as

$$I_x = J_x \cdot (W \cdot d) = p \cdot q \cdot v_x (W \cdot d) \tag{1.51}$$

where

W	Width.
d	Depth with cross-sectional area (W · d) for the current.
$p = N_A$	Hole carrier density.

On putting v_x of Eq. (1.50) in Eq. (1.51), we get

$$I_x = p \cdot q \left(\frac{V_{Hy}}{B_z \cdot W} \right) (W \cdot d) \tag{1.52}$$

$$\therefore \text{Hall voltage} \quad V_{Hy} = \left(\frac{I_x B_z}{p \cdot q \cdot d} \right) = \frac{I_x B_z}{d} \cdot R_H \tag{1.53}$$

where

$R_H = (1/qp) = $ Hall coefficient.

From Eq. (1.52),

$$p = \left(\frac{I_x B_z}{V_{Hy} \cdot q \cdot d} \right) \tag{1.54}$$

Thus, we get the carrier density and hence doping (N_A) in the sample, if Hall voltage and current are known. Similarly, for *n*-type semiconductor, where the Hall voltage is −ve, we get

$$n = - \left(\frac{I_x B_z}{q d V_{Hy}} \right) \tag{1.55}$$

For computing mobility, we write the current density equation with field $E_x = V_x/L$

$$J_{xp} = q \cdot p \cdot \mu_p E_x = q \cdot p \cdot \mu_p \frac{V_x}{L}$$

$$\therefore \quad \text{Current } I_{xp} = J_{xp} \cdot W \cdot d = q p \mu_p \frac{V_x}{L} \cdot W \cdot d \tag{1.56}$$

$$\therefore \quad \mu_p = \frac{I_{xp} \cdot L}{q \cdot p \cdot V_x \cdot W \cdot d}$$

Thus, we get the hole mobility in *p*-type semiconductor, if the doping density *p* and Hall Voltage are known.

Similarly, electron mobility by using *n*-type semiconductor will be

$$\mu_n = \frac{I_{xn} \cdot L}{q \cdot n \cdot V_x \cdot W \cdot d} \tag{1.57}$$

where *n* and *p* are given by Eq. (1.55) and Eq. (1.54), respectively.

1.13   Comparison of Properties

1.13.1   Comparison Between Metals (Conductors) and Semiconductors

Metals (Conductors)	Semiconductors
1. Charge carriers are only electrons	1. Charge carriers are electrons and holes
2. Low resistivity or high conductivity	2. Moderate resistivity and conductivity
3. Valence band and conduction band overlap or a part of conduction band lies in valence band and therefore $E_G = 0$	3. Valence band and conduction band are separated by a small energy gap (<3 eV)
4. Resistance increases with rise in temperature, i.e. metals have positive temperature coefficient of resistance.	4. Resistance decreases with rise in temperature, i.e. metals have negative temperature coefficient of resistance.
5. High carrier, density of the order of 10^{22} to 10^{23}/cc	5. Intrinsic carrier density (At 300 K, for Si, $n = 1.5 \times 10^{10}$/cc and for Ge, $n = 2.5 \times 10^{13}$/cc) which increases with doping and temperature
6. Current flows only by Drift process	6. Current flows due to drift and diffusion
7. Conductivity is given as $\sigma = nq\mu$	7. Conductivity in semiconductors is given by $\sigma = q(n\mu_n + p\mu_p)$
8. Resistivity of metals increases with increasing impurity concentration	8. Resistivity of semiconductors decreases with increasing impurity concentration
9. The optical radiation does not affect the resistivity of metals	9. Resistivity of semiconductors decreases with increases in light intensity/some radiation, e.g. X-ray, etc.

1.13.2   Comparison Between Intrinsic and Extrinsic Semiconductors

Intrinsic Semiconductor	Extrinsic Semiconductors
1. These are pure semiconductors (no impurities or lattice defects)	1. These are impure semiconductors (impurities) of pentavalent or trivalent which are added in a small amount, i.e. 0.0001%
2. Conductivity of intrinsic semiconductor is poor	2. Conductivity of extrinsic semiconductor is large
3. Here the number of electrons and holes is equal	3. Here either majority are electrons for n-type or majority are holes for p-type semiconductor
4. Fermi level lies at the centre of forbidden gap	4. Fermi level lies near the bottom of conduction band for n-type or near the top of valence band for p-type

1.13.3   Comparison Between n-Type and p-Type Semiconductors

n-Type	p-Type
1. Here, pentavalent impurities like As, Sb, Bi, etc. are added	1. Here, trivalent impurities like In, Al, B, Ga, etc. are added
2. Also called donor-type impure semiconductors	2. Also called acceptor-type impure semiconductors
3. Here, electrons are the majority carriers and holes are the minority carriers	3. Here, holes are the majority carriers and electrons are the minority carriers
4. Here $N_A = 0$ and $n \gg p$; $n = N_D$, and $p = n_i^2/N_D$	4. Here $N_D = 0$ and $p \gg n$; $p = N_A$ and $n = n_i^2/N_A$
5. For n-type semiconductor, Fermi level and donor level lie near the bottom of conduction band at 0 °K	5. For p-type semiconductor, Fermi level and acceptor level lie near the top of valence band at 0 °K

Table 1.4 Electrical properties of Ge and Si

S. No.	Property	Ge	Si
1.	Atomic number	32	14
2.	Atomic weight	72.6	28.1
3.	Relative dielectric constant	16	12
4.	Atoms/cm^3	4.4×10^{22}	5.0×10^{22}
5.	E_{g0} (eV) at 0 °K	0.785	1.21
6.	E_g (eV) at 300 °K	0.72	1.10
7.	n_i at 300 °K, cm^{-3}	2.5×10^{13}	1.5×10^{10}
8.	Intrinsic resistivity at 300 °K, Ω-cm	45	2,30,000
9.	μ_n cm^2/V-s at 300 °K	3,800	1,300
10.	μ_p cm^2/V-s at 300 °K	1,800	500
11.	D_n cm^2/s = $\mu_n V_T$	99	34
12.	D_p cm^2/s = $\mu_p V_T$	47	13
13.	Maximum saturation velocity of electron v_S (cm/s)		10^7

1.13.4 Electrical Properties of Ge and Si

The electrical properties of germanium and silicon are listed in Table 1.4.

Questions

Fill up the blanks:

1. At room temperature, the current in an intrinsic semiconductor is due to _____.
2. An intrinsic semiconductor at the absolute zero temperature behaves like _____.
3. The breakdown mechanism in a lightly doped *pn* junction under reverse-biased condition is called _____.
4. _____ diode is operated in reverse bias mode.
5. Space charge region around a *pn* junction does not contain _____.
6. Diode is a/an _____ device.
7. *n*-type semiconductor are superior to *p*-type semiconductor, because mobility of electrons is _____ than that of holes.
8. *n*-type silicon is obtained by doping with _____ element and *p*-type silicon is obtained by doping with _____ element.
9. When the temperature of a doped semiconductor is increased, its conductivity _____.
10. A Zener diode has a sharp breakdown at low _____.

Short Questions

1. Give the energy band structure of insulators, semiconductors and conductor.
2. Explain the effect of temperature on semiconductors.
3. Briefly explain *n*-type and *p*-type semiconductors.
4. Define the terms conductivity and mobility in a semiconductor and why we prefer Si over Ge (Germanium).
5. What are the majority and minority carriers for both *p*-type and *n*-type semiconductors and explain how to form *pn* junction diode?
6. Write the application of *pn* junction diode and give the example for the application of Zener diode.
7. Give short notes about the *V–I* characteristics of *pn* junction diode.
8. Give short notes about the *V–I* characteristics of Zener diode.
9. Classify materials on the basis of electrical conductivity.
10. Distinguish between free electrons and holes.
11. What do you understand by intrinsic and extrinsic semiconductors?
12. What is Zener diode? Why is Zener diode used as voltage regulator?
13. Define depletion layer. Give the structure.
14. Define Fermi level and forbidden energy gap.
15. Distinguish between drift and diffusion current.

Long Questions

1. Explain the operation of *pn* junction diode.
2. Draw the V-I characteristics of *pn* junction diode.
3. Explain the Zener diode and draw its characteristics under forward and reverse bias.
4. Differentiate between Zener breakdown and Avalanche breakdown.
5. Differentiate between intrinsic and extrinsic semiconductors.
6. Explain the action of *pn* junction diode under forward bias and reverse bias. Write down the application of *pn* junction diode.
7. Differentiate between the *p*-type and *n*-type semiconductors.

Chapter 2
Semiconductor Diodes and Application

Contents

© Springer Nature Singapore Pte Ltd. 2020
S. S. Srikant and P. K. Chaturvedi, *Basic Electronics Engineering*,
https://doi.org/10.1007/978-981-13-7414-2_2

2.1   The *pn* Junction Diode

The semiconductor was discovered in 1931 (Wilson) followed by the step of adding *p*- and *n*-type impurities in it. As such these *n*- or *p*-type semiconductor materials' individuals had very limited application, which is true today also. The real fast development started from 1948 (Shockley) after the understanding of the mechanism which comes into play when a *p*-type semiconductor is paired with *n*-type semiconductor. Almost all useful devices/applications involve using this versatile *pn* junction or a combination of such junctions. A *pn* junction cannot be created by joining a *p*-type with an *n*-type semiconductors. Its fabrication is a process called *n*-type of impurity diffusion (higher than in *p*-type) into a *p*-type silicon wafer or the other way round.

Historically alloyed junction was first made, where a *p*-type impurity (Al metal) was put on an *n*-type silicon and heated to eutectic temperature (≈ 580 °C) for Al–Si system. This leads to the formation of puddle of molten Al–Si mixture as shown in Fig. 2.1a. This after cooling results in recrystalized Si-saturated (p^+) with *p*-type acceptor impurity. Then Al on the bottom of the silicon in evaporated for metallic ohmic contact. Thus, we used to get a *pn* junction.

Today, the method is different and is by diffusion of impurity into Si-wafer. Here, a trivalent (i.e. *p*-type impurity) as a gaseous compound (e.g. AsH_3) is passed on a lightly doped *n*-type silicon wafer at 1050 °C. This impurity (e.g. As) goes into the Si by diffusion leaving the hydrogen gas and making a layer of *p*-type on the silicon wafer as shown in Fig. 2.1b. Therefore, metallization of top and bottom surfaces is done for wire contacts.

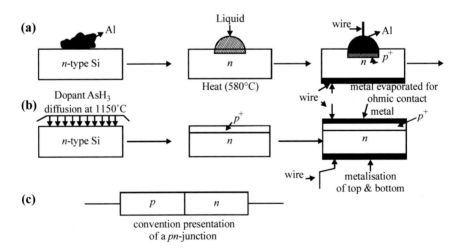

Fig. 2.1 Formation of *pn* junction **a** Alloy junction method, **b** impurity diffusion method, **c** conventional presentation of a *pn* junction

In fact, the process of fabrication *pn* junction diodes and other devices has further advanced called planar technology. In this technology, all the metal contacts for *p* and *n* sides are taken from the top surface only, with all the processes of diffusion, etc. done from the upper plane only.

2.1.1 Depletion Region Formation in pn Diode Without Bias

Figure 2.2a shows the *pn* junction diode just before diffusion of charge carriers, while Fig. 2.2b shows the formation of depletion region after diffusion of carriers without bias. The cause of diffusion is due to the fact that *p* side has excess holes and n side has excess electrons (Fig. 2.2a). Each of the excess charges diffuse to the other side across the junction due to attraction (Fig. 2.2b). This leads to the formation of electric field and voltage [called Barrier Potential (V_{BP})] along BJA (Fig. 2.2b). This field stops further diffusion of the two charge forming depleted regions AJ and JB across J (Fig. 2.3) which then reach equilibrium. Finally, CA region contains holes as majority plus electrons as minority, while AJ region is left with immobile −ve acceptor ions. Similarly, DB region contains electrons as majority carriers plus holes as minority carriers, while BJ region only +ve immobile donor ions. Thus AB region (Fig. 2.2b) becomes depletion region depleted of mobile charges. The sample as a whole remains neutral.

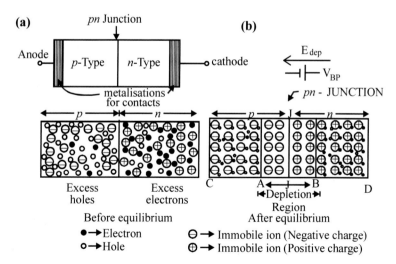

Fig. 2.2 a *pn* junction just before diffusion of, **b** *pn* junction just after diffusion of carriers forming a depletion region (AB) void of mobile carriers

Fig. 2.3 *pn* junction without bias **a** charge density concentration, **b** electric field, **c** barrier potential, **d** energy level in *pn* junction band diagram

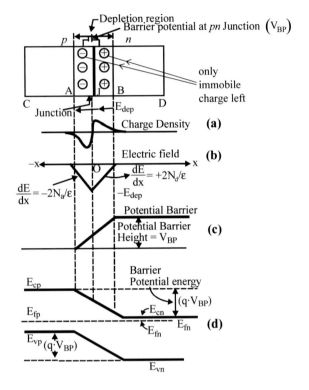

Thus once the *pn* junction is formed, following action is summarized:

1. Due to difference of concentrations of holes and electrons in *p* and *n* sides and also due to opposite charges of these mobile charges, diffusion takes place. The thermal energy agitated in random direction helps the diffusion process as shown in Fig. 2.2b.

2. The electrons of *JB* region of *n* side diffuse and combine with the free holes in *JA* region and neutralize them. Similarly, the free holes of *JA* region diffuse into the *JB* region of *n* side and neutralize them.

3. This makes the *AB* region devoid (depleted) of mobile charges and is left with charged immobile −ve acceptor ions in *JA* side and +ve donor ions in *JB* side. This charge devoid region (*AB*) is depletion region and therefore it is of high resistance region. This depletion region (or transition region) is insulator like and very thin <1 μm or so.

4. This process of diffusion continues only for a very short time, reaching an equilibrium due to the restraining force of electric field (E_{dep}) automatically set up due to +ve charge in *JB* and −ve charge in *JA*. Naturally, the direction of this electric field (E_{dep}) is from *JB* region to *JA* region, as shown in Fig. 2.3b.

5. Corresponding to this electric field (E_{dep}) and depletion width (*W*), the voltage will be there which is called barrier potential (V_{BP}) as it becomes the barrier for further diffusion. This $V_{BP} = 0.3$ V for Ge and 0.7 V for Si. **This barrier**

potential (V_{BP}) is also called built-in voltage, having got built due to diffusion as shown in Fig. 2.3c.

In the energy band diagram, both E_{cp} and E_{cn} energy levels get separated (p side moving up and n side moving down) by barrier potential energy of $(q \cdot V_{BP})$, with Fermi levels unshifted ($E_{fp} = E_{fn}$) both sides as shown in Fig. 2.3d.

Depletion layer: A region around the junction from which the charge carriers (free of mobile electrons and holes) are depleted is called depletion layer.

Depletion region = Transition region = Space Charge Region

In an unbiased *pn* junction, the depletion region get formed immediately after the formation of the junction.

2.1.2 Reverse Biasing and Minority Current

When the positive terminal of a *dc* source or battery is connected to *n*-type and negative terminal is connected to *p*-type semiconductor material of a *pn* junction, then the junction is called 'Reverse biased'.

As clearly shown in Fig. 2.4A, the applied reverse potential (V_R) acts in such a manner that it establishes an electric field in the same direction as the potential barrier (V_{BP}) sets up without any bias, as shown in Fig. 2.3. Therefore, this reverse bias adds to the potential barrier and increases the depletion width. The increased potential barrier prevents the flow by majority carriers across the junction. However, minority current flows which is very small of the order of µA and is leakage in nature. Thus, one can say that there is a high resistive path, established by the junction, which is of the order of kΩ (as R_R shown in Fig. 2.6). In the energy band diagram, the energy levels E_{cp}, E_{cn} get separated by $q \cdot (V_{BP} + V_R)$, with Fermi levels E_{fp}, E_{fn} by $(q \cdot V_R)$ only.

2.1.3 Forward Biasing and Majority Current

When the positive terminal of a *dc* source or battery is connected to *p*-type and negative terminal is connected to *n*-type semiconductor of a *pn* junction, then the junction is said to be in forward biasing. It is shown in Fig. 2.4B.

In this case, the applied forward potential acts in such a way that it established an electric field opposite to that and reduces the field due to potential barrier (V_{BP}). Thus, the barrier potential at the junction is reduced. Since the potential barrier voltage (V_{BP}) is very small (0.7 V for silicon and 0.3 V for germanium junction), **therefore forward bias first cancels this built-in voltage, i.e. barrier voltage (V_{BP}) and then only by ($V_F - V_{BP}$) voltage, current can flow.** Therefore, the forward turn-on voltage/knee voltage as shown in *I-V* characteristic (Fig. 2.7) is equal to V_{BP}.

Fig. 2.4A *pn* junction in a
reverse biasing (V_R):
a depletion region, **b** field
inside and **c** energy level in *pn*
junction band diagram.

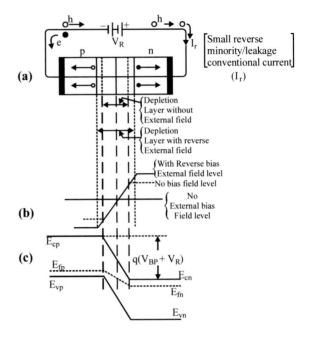

Fig. 2.4B *pn* junction under
forward biasing (V_F):
a depletion region, **b** field
inside it and **c** energy levels in
pn junction band diagram

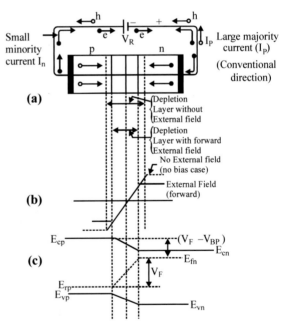

Therefore, once the potential barrier is cancelled/neutralized by the forward voltage, a conducting path is established for the flow of current. Thus, a large current starts flowing through the junction. This current is called forward current, which is of order of mA.

Hence, the external voltage applied to a *pn* junction, which cancels the potential barrier to constitute easy flow of current through it, is called forward biasing.

The following points are worth noting, when a *pn* junction is forward biased:

(i) Holes are pushed from the positive terminal of the battery and forced towards the junction.

(ii) Similarly, electrons are pushed from the negative terminal of the battery and forced towards the junction.

(iii) Because of this increased energy, some holes and electrons enter the depletion region. This cancels the potential barrier. Therefore, width of the depletion region also reduces.

(iv) More majority carriers diffuse across the junction. Hence, majority carriers cause a large current to flow across the junction of the order of mA.

(v) The junction offers low resistance (forward resistance, R_F) to the flow of current through it, because of large number of carriers present in whole of the diode. This R_F is of the order of a few unit ohms (as shown in Fig. 2.6 in Art 2.1.4).

Summary: It is concluded from the above discussion that with forward biasing, a low resistance path ($R_F \approx$ units of Ω) is set up by the *pn* junction and hence current flows through the circuit because of majority charge carriers. On the other hand, the minority current flows in reverse biased, which provides a high resistance path (R_R of the order of $k\Omega$) by the *pn* junction.

2.1.4 Total V-I Characteristics of a pn Junction Diode—Experimentally

The total *V-I* (volt-ampere) characteristics of a *pn* junction is the curve between voltage across the junction and the current through it. Figure 2.5 shows the circuits arrangement to determine the V-I characteristics of a *pn* junction.

With the help of solid-state physics, the diode current equation, relating the voltage *V* and current *I* for the forward- and reverse-bias regions, is given by

$$I = I_0 \left(e^{V/\eta V_T} - 1 \right) \tag{2.1}$$

where

I = Diode current,
I_0 = Diode reverse saturation current at room temperature,
η = Constant [$\eta = 1$ for Ge, $\eta = 2$ for Si],

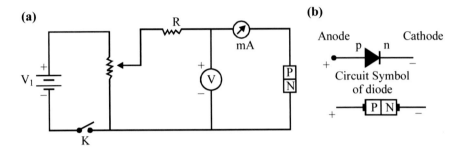

Fig. 2.5 **a** Circuit arrangement of *pn* junction for getting V-I characteristics, **b** circuit symbol of an ideal diode. Here, external voltage V_1 has to be +ve for forward bias and −ve for reverse bias

Fig. 2.6 V-I characteristics of *pn* junction

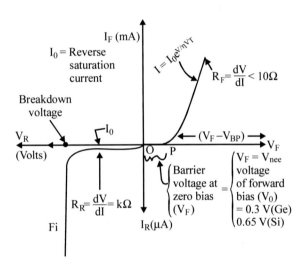

V = External voltage applied to the diode,

$V_T = \frac{kT}{q} = \frac{T}{11600}$ = Voltage equivalent of temperature (Thermal voltage = 0.026 V),

k = Boltzmann's constant,

= 1.38066 × 10⁻²³ J/K,

q = Electronic charge = 1.6 × 10⁻¹⁹ C,

T = Diode junction temperature in °K, and

At the temperature, i.e. T = 300 °K, VT = 26 mV.

Thus, the characteristics of *pn* junction diode as shown in Fig. 2.6 can be studied by visualization of the behaviour of *pn* junction under following conditions:

1. **Zero external voltage**: When the circuit is open, i.e. the external voltage is zero, the diffusion of charges across the junction builds in a voltage (V_{BP}), which acts as the potential barrier, not allowing the current to flow. Therefore, the circuit current is zero as shown by point '*O*' in Fig. 2.6.

2. **Reverse Bias**: With a reverse bias to the *pn* junction diode, the potential barrier is increased. Therefore, the junction resistance becomes very high and a very small current flows in the circuit with reverse bias known as reverse saturation current ($I_0 \approx \mu A$). On continuously increasing the reverse voltage, breakdown of junction takes place, where there is a sudden rise of reverse current, which is controlled by external series resistance if connected or else it get burned.

From Eq. 2.1 for reverse-biased condition, $(V = -ve)$ Therefore

$$e^{V/\eta V_T} \ggg 1$$
$$\text{So,} \quad e^{V/\eta V_T} - 1 \approx -1$$
$$\therefore \quad I_R = I_0\left(e^{V/\eta V_T} - 1\right) \approx -I_0$$

$$I_R \approx -I_0 \qquad\qquad\qquad ...(2.2)$$

This equation is valid as long as the external voltage is below the breakdown value. (see Fig 2.6)

3. **Forward Bias**: With forward bias to the *pn* junction, the potential barrier is reduced. After knee voltage ($=V_{BP}$), the current increases ($\approx mA$) exponentially with the increase in forward bias and the curve is therefore nonlinear.

From Eq. 2.1, for the forward-biased condition ($V = +ve$):

$$e^{V/\eta V_T} \gg 1$$
$$\text{So} \quad e^{V/\eta V_T} - 1 \approx e^{V/\eta V_T} \qquad\qquad (2.3)$$
$$\therefore \quad I_F = I_0\left(e^{V/\eta V_T} - 1\right) \approx I_0 e^{V/\eta V_T}$$

2.1.5 Dependence of Reverse Leakage Current (I_0) on Temperature

There are some important factors related to the reverse saturation current (I_0):

(i) The reverse saturation current (I_0) increases at a rate of 7% for every 1 °C rise in temperature.

(ii) The reverse saturation current (I_0) doubles its value for every 10 °C rise in temperature.

(iii) The increasing levels of I_0 with temperature account for the lower levels of cut-in voltage (Fig. 2.7).

The typical values of reverse saturation current (I_0) for silicon are much lower than that of germanium for similar temperature, power, and current levels. Due to this reason, the levels of I_0 in silicon even at high temperature also remain lower than in germanium. **This is also one of the main reasons that silicon devices have reached a significantly higher level of development in design and application.**

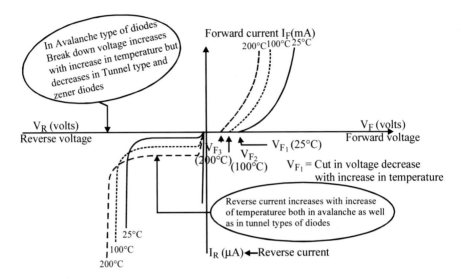

Fig. 2.7 Effects of temperature on V-I characteristics of a silicon diode for avalanche type of breakdown

(iv)   At $I_0 = I_{01}$, for temperature T_1, the value of reverse saturation current (I_0) at temperature (T) is given by

$$I_0(T) = I_{01} \times 2^{(T-T_1)/10} \tag{2.4}$$

The effects of temperature on V-I characteristics of a silicon semiconductor is shown in Fig. 2.7.

2.1.6   Calculations of Built-in Potential (V_{BP}) and Depletion Layer Width, with or Without Bias

(a)   **Built-in Potential or Potential barrier**: As shown in Chap. 1 that for p-type and n-type semiconductors, the Fermi level are

$$E_{fp} = E_{ip} - kT \ \ln(N_A/n_i) \tag{2.5a}$$

$$E_{fn} = E_{in} + kT \ \ln(N_D/n_i) \tag{2.5b}$$

where

E_{ip}, E_{in} = Fermi levels of intrinsic semiconductor in p side and n side, respectively, which are the midpoints of E_C and E_V at each sides separately.

N_A, N_D = Acceptor and donor impurity densities in p side and n side, respectively.

n_i = Intrinsic carrier concentration of the semiconductor.

In a *pn* junction, n and p materials are continuous, so their Fermi levels have to be equal, i.e. $E_{fp} = E_{fn}$; therefore, equating the above two equations, we get the separation of the intrinsic Fermi levels of n and p sides as

$$\left(E_{ip} - E_{in}\right) = kT \ln\left(\frac{N_D}{n_i}\right) + kT \ln\left(\frac{N_A}{n_i}\right) = kT \ln\left(\frac{N_A N_D}{n_i^2}\right) \qquad (2.5c)$$

As the energy band diagrams of n and p side have shifted from each other by built-in potential energy, i.e. the potential barrier energy ($=q\ V_{BP}$), therefore their midpoints, i.e. E_{ip} and E_{in} also got shifted by the same amount.

$$\therefore \quad \left(E_{ip} - E_{in}\right) = q V_{BP} = kT \ln\left(N_A \cdot N_D/n_i^2\right) \qquad (2.5d)$$

$$\therefore \quad \text{Build in Potential} = \text{Barrier Potential} = V_{BP} = \frac{kT}{q}\ln\left(N_A \cdot N_D/n_i^2\right)$$

$$(2.5e)$$

Thus, the value of V_{BP} depends upon the doping of n, p sides and temperature of a given material, as temperature controls n_i by Eq. 1.3 (Fig. 1.13).

As per Fig. 2.7, the turn-on or knee voltage is equal to V_{BP}; therefore, by the *I*-V characteristics, we can get the value of V_{BP}. Therefore, if N_A is known, we can compute N_D from Eq. (2.5e) of V_{BP}.

(b) **Depletion layer width and maximum electric field without bias (in abrupt *pn* junction)**: We know that after diffusion of charge carriers across the *pn*, electric field is built in, which begin from +ve charge and end on −ve charge. This space charge extend unequally into the unequally doped p and y regions (x_p and x_n) of the depletion region and depends upon their doping concentrations, for uncovering equal amount of charges, i.e. $q\ N_D\ x_n = q\ N_A\ x_p$; therefore, $\frac{x_n}{x_p} = \frac{N_A}{N_D}$ (see in Fig. 2.8). For example, if n side is less doped as compared to the p side, then $x_x > x_p$. Same is true for the built-in voltages of the two sides ($V_n > V_p$), with total built-in potential barrier as $V_{BP} = (V_n + V_p)$.

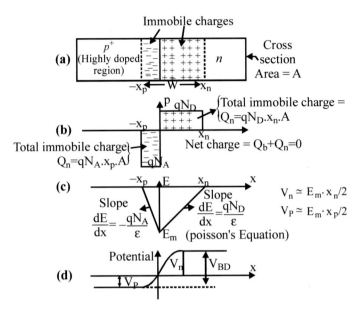

Fig. 2.8 **a** *pn* junction with p side heavily doped, i.e. p$^+$ with space charge region from $-x_n$ to $+x_n$, **b** charge density distributions as rectangular, **c** electric field distributions, **d** potential variation and Barrier potential

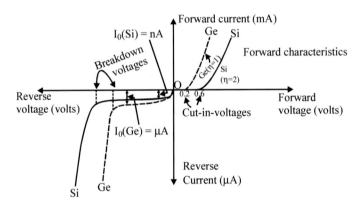

Fig. 2.9 Comparison of V-I characteristics of silicon and germanium diode

For convenience, we take that each of the *n* and *p* depletion regions have uniform charge density distributions along their lengths (x_n, x_p) as shown in Fig. 2.8, as a result the areas of these two charge regions will be equal as

$$q \cdot N_D \cdot x_n = q \cdot N_A \cdot x_p. \tag{2.6a}$$

Also, the electric field in whole of the depletion region is same therefore

$$E = \frac{V_n}{x_n} = \frac{V_P}{x_P} \tag{2.6b}$$

Therefore, by Eqs. (2.6a) and (2.6b), we get

$$\frac{V_n}{V_P} = \frac{x_n}{x_P} = \frac{N_A}{N_D} \tag{2.6c}$$

If we add 1 to Eqs. (2.6a, 2.6b, 2.6c), it gives us depletion width W and barrier voltage V_{BP}

$$\text{As} \quad W = x_n + x_P \text{ and } V_{BP} = V_n + V_P$$

$$\therefore \quad \frac{V_n + V_P}{V_P} = \frac{x_n + x_P}{x_P} = \frac{N_A + N_D}{N_D}$$

$$\therefore \quad x_P = \frac{(x_n + x_P) \cdot N_D}{N_A + N_D} = \frac{W \cdot N_D}{N_A + N_D} \tag{2.7a}$$

$$V_P = \frac{V_{BP} \cdot N_D}{N_A + N_D} \tag{2.7b}$$

$$\text{Similarly} \quad x_n = \frac{W \cdot N_A}{N_A + N_D} \tag{2.7c}$$

$$V_n = \frac{V_{BP} \cdot N_D}{N_A + N_D} \tag{2.7d}$$

The Poisson equations for the p and n sides due to the immobile charges will be, respectively,

$$\text{and} \quad \left. \begin{array}{l} \frac{dE}{dx} = -q \frac{N_A}{\varepsilon} \\ \frac{dE}{dx} = +q \frac{N_D}{\varepsilon} \end{array} \right\} \tag{2.8}$$

Integrating them from $x = -x_p$ to 0; $E = 0$ to E_m and $x = 0$ to x_n; $E = E_m$ to 0, respectively (where E_m is the maximum electric field), we get

$$E_P(x) = +\frac{qN_A}{\epsilon}(x + x_p) = E_m\left(1 + \frac{x}{x_p}\right) \tag{2.9a}$$

and

$$E_n(x) = -\frac{qN_D}{\epsilon}(x - x_n) = E_m\left(1 - \frac{x}{x_n}\right) \tag{2.9b}$$

where $E = E_m = q\frac{N_A x_p}{\epsilon}$ for p side and $E = E_m = q\frac{N_D x_n}{\epsilon}$ for n side at $x = 0$. At the junction, maximum field is same for both sides therefore $E_m = q\frac{N_A x_p}{\epsilon} = q\frac{N_D x_n}{\epsilon}$.

As electric field is given by $E = -dV/dx$, therefore integrating the field for the whole region $(-x_P$ to $+x_n)$, built-in potential will be

$$V_{BP} = -\int_{x_p}^{x_n} E \cdot dx = -\int_{-x_p}^{0} E \cdot dx - \int_{0}^{x_n} E \cdot dx$$

$$V_{BP} = \left(\frac{qN_A x_p^2}{2\epsilon} + \frac{qN_D x_n^2}{2\epsilon}\right) = \frac{1}{2}E_m(x_p + x_n) = \frac{1}{2}E_m.W \tag{2.10}$$

On putting the values of x_P and x_n of Eqs. (2.7a) and (2.7c) in the first part of Eq. (2.10), we get

$$V_{BP} = \frac{q}{2\epsilon}\left[N_A\left(\frac{N_D W}{N_A + N_D}\right)^2 + N_D\left(\frac{N_A W}{N_A + N_D}\right)^2\right]$$

$$V_{BP} = \frac{q}{2\epsilon}W^2\frac{N_A N_D}{N_A + N_D} \tag{2.11}$$

$$\therefore \quad \text{Depletion Width } (W) = \sqrt{\frac{2\epsilon}{q}V_{BP}\left(\frac{1}{N_A} + \frac{1}{N_D}\right)}$$

For one-sided abrupt junction, i.e. with $N_A \gg N_D$ is $(1/N_A) \ll (1/N_D)$ the depletion width (W) will be

$$W = \sqrt{\frac{2\epsilon V_{BP}}{q N_D}} \tag{2.11a}$$

This shows that most of the portion of the depletion region as well as the built-in potential or barrier potential (V_{BP}) is in the lower doped side only, for one-sided abrupt junction.

> Abrupt junction is that specially made *pn* junction where doping level changes abruptly or very sharply (not gradually) across the junction, with very very high doping on one side.

(c) **Depletion layer width and maximum electric field (E_m) with bias**: As is clear that the reverse bias (V_r) increases the depletion width W, as more mobile charges will move away towards the p and n sides, increasing the peak field E_m as well (Eq. 2.10). This is because V_r gets added to V_{BP} and therefore opposite is true for forward bias in Eq. 2.11 and hence in Eq. 2.11a. Here, we define $V = V_f$ for forward bias and $V = -V_r$ for reverse bias.

Example 2.1 The forward current through a silicon diode is 10 mA at room temperature (27 °C). The corresponding forward voltage is 0.75 V. Find the reverse saturation current (I_0).

Solution

$$I_F = 10\,\text{mA}, V_F = 0.75\,\text{V}$$
$$T = 273 + 27 = 300\,°\text{K}$$
$$\eta = 2 \text{ for Si diode}$$
$$V_T = \frac{T}{11600} = \frac{300}{11600} = 0.026\,\text{V} = 26\,\text{mV}$$

The forward current through a diode is given by

$$I_F \approx I_0 e^{V_F/\eta V_T}$$
$$10 \times 10^{-3} \approx I_0 \exp\left(\frac{0.75}{2 \times 26 \times 10^{-3}}\right)$$
$$I_0 \approx 5.4 \times 10^{-9}\text{A}$$
$$I_0 = 5.4\,\text{nA}$$

This being very small and hence it is sometimes called leakage current.

Example 2.2 A germanium diode carries a current of 1 mA of room temperature, when a forward bias of 0.15 V is applied. Calculate the reverse saturation current at room temperature **[UPTECH 2005–2006, First Semester]**.

Solution Given

$$T = 273 + 27 = 300\,°\text{K}$$
$$V_F = 0.15\text{V}, I_F = 1\,\text{mA} = 1 \times 10^{-3}\text{A}$$
$$\eta = 1 \text{ for Ge}, V_T = \tfrac{T}{11600} = 0.026\,\text{V}$$

∴ For forward-bias current

$$I_F \approx I_0 \exp\left(\frac{V_F}{\eta V_T}\right)$$

$$1 \times 10^{-3} \approx I_0 \exp\left(\frac{0.15}{1 \times 0.026}\right)$$

$$I_0 \approx 3.1 \times 10^{-6} = 3.1\,\mu A \text{ (Much larger than in } Si)$$

Example 2.3 Derive an expression for the ratio of diode currents resulting from two different forward voltage (V) if $V \gg \eta V_T$.

Solution If $V \gg \eta V_T$, then diode equation becomes

$$I = I_0 e^{V/\eta V_T}$$
$$\therefore \quad \frac{I_2}{I_1} = \frac{I_0 e^{V_2/\eta V_T}}{I_0 e^{V_1/\eta V_T}} = e^{(V_2 - V_1)/\eta V_T}$$

Example 2.4 If the current of a silicon diode with $V_T = 26$ mV doubles by increasing the forward bias, find this increase.

Solution As we know that

$$\frac{I_2}{I_1} = e^{(V_2 - V_2)/\eta V_T} = e^{\Delta V/\eta V_T}$$

$$\frac{I_2}{I_1} = 2, \eta = 2 \text{ for } Si$$

$$V_T = 0.026 \text{ V}$$

$$\therefore \quad 2 = \exp\left(\frac{\Delta V}{2 \times 0.026}\right)$$

$$In(2) = \left(\frac{\Delta V}{2 \times 0.026}\right)$$

$$\Delta V = 2 \times 0.026 \times \ln 2 = 36.04 \text{ mV}$$

Example 2.5 A germanium diode at $T = 300$ °K conducts 5 mA at $V = 0.35$ V. Find the diode current if $V = 0.4$ V.

Solution

$$I_1 = 5\,\text{mA}, V_1 = 0.35\,\text{V}$$
$$I_2 = ?, V_2 = 0.40\,\text{V}$$
$$\eta = 1 \text{ for } Ge; V_T = \frac{T}{11600} = 26\,\text{mV}$$

$$\therefore \quad \frac{I_2}{I_1} = e^{(V_2 - V_1)/\eta V_T}$$

$$I_2 = I_1 e^{(0.4 - 0.35)/(1 \times 26 \times 10^{-3})}$$

$$I_2 = 34.6 \, \text{mA}$$

i.e. just by increasing forward bias by 0.05 V current has got increased from 5 to 34.6 mA, i.e. 7 times or 80.

Example 2.6 Calculate the built-in potential for Ge, Si and GaAs at 300 °K. Given

$$n_i = 2.4 \times 10^{13}/\text{cc}(\text{Ge}); \; 1.5 \times 10^{10}/\text{cc}(Si); \; 1.79 \times 10^{6}/\text{cc}(\text{GaAs})$$

The doping density of *p* side and *n* side are as 10^{18}/cc and 10^{15}/cc, respectively.

Solution

$$N_D = 10^{18}/\text{cc}; \qquad N_A = 10^{15}/\text{cc}$$
$$V_{BP} = \frac{kT}{q} \ln\left(\frac{N_A N_D}{n_i^2}\right); \quad \frac{kT}{q} = 0.026 \, \text{V} \, at \, 300°\text{K}$$

For Ge,

$$V_{BP} = 0.026 \, \ln\left(\frac{10^{18} \times 10^{15}}{(2.4 \times 10^{13})^2}\right) = 0.37 \, \text{V}$$

For Si,

$$V_{BP} = 0.026 \, \ln\left(\frac{10^{18} \times 10^{15}}{(10^5 \times 10^{10})^2}\right) = 0.73 \, \text{V}$$

For GaAs,

$$V_{BP} = 0.026 \, \ln\left(\frac{10^{18} \times 10^{15}}{(1.79 \times 10^6)^2}\right) = 1.23 \, \text{V}$$

Example 2.7 Calculate the built-in potential, W_{dep} and E_m at zero bias for an abrupt junction with $N_A = 10^{19}$/cc; $N_D = 10^{15}$/cc at room temperature for Si.

Solution For Si, at room temperature (300 °K),

$$n_i = 1.5 \times 10^{10}/\text{cc}, \epsilon_r = 11.9;$$
$$\epsilon_0 = 8.85 \times 10^{-14} \, \text{F/cm}, \epsilon = \epsilon_r \epsilon_0$$

Given

$$N_D = 10^{15}/\text{cc}, N_A = 10^{19}/\text{cc}; \frac{kT}{q} = 0.026\text{V} \text{ at } 300\,°\text{K}$$

$$V_{BP} = \frac{kT}{q}\ln\left(\frac{N_A N_D}{n_i^2}\right) = 0.026 \times \ln\left[\frac{10^{15} \times 10^{19}}{(1.5 \times 10^{10})^2}\right]$$

Built-in potential = V_{BP} = 0.817 V.

This is quite high as doping in both sides are high.
Depletion width

$$W = \sqrt{\frac{2 \in V_{BP}}{qN_D}} = \sqrt{\frac{2 \times 11.9 \times 8.854 \times 10^{-14} \times 0.817}{1.6 \times 10^{-19} \times 10^{15}}} \approx 1.037 \ \mu\text{m}$$

$$E_m = \frac{2V_{BD}}{W} = \text{ Maximum Electric field for an abrupt junction}$$

$$E_m = \frac{2 \times 0.817}{1.037 \times 10^{-4}} = 1.6 \times 10^4 \text{ V/cm}$$

2.1.7 Comparison of Silicon and Germanium Diodes

S. no.	Parameters	Silicon diode	Germanium diode
1.	Material used	Silicon	Germanium
2.	Diode equation $I = I_0 \left[e^{(V/\eta kT)} - 1\right]$	Here $\eta = 2$	Here $\eta = 1$
3.	Cut-in voltage (or turn-on voltage) [depends on doping also]	≈0.6 V	≈0.2 V
4.	Effect of temperature	Less	More
5.	Reverse saturation current (I_0)	In nanoampere	In microampere
6.	Breakdown voltage	Higher	Lower
7.	Applications	Rectifier, clipper, dampers, etc.	Low-voltage, low-temperature applications
8.	V-I characteristics as shown in Fig. 2.9	Represented by lines	Represented by lines

2.2 Breakdown Diodes—Avalanche and Zener

When in normal diodes reverse-bias 'breakdown' takes place and a large current flows in reverse direction, it can cause damage to the diode. However, it is possible to build a special type of diode (Avalanche and Zener) that can be operated in a breakdown region, under controlled current condition. This section is devoted to such type of diodes that are designed to operate in the breakdown region.

2.2.1 Breakdown Mechanisms

If the reverse bias applied to the *pn* junction is increased, a point will reach when the junction breaks down and reverse current rises sharply to a value limited only by the external resistance connected in series. This specific value of the reverse-bias voltage is called breakdown voltage. After breakdown, a very small increase of reverse voltage increases the reverse current very much. The breakdown voltage depends upon the width of depletion layer, which in turn depends upon the doping level.

The phenomenon of diode breakdown is due to two different types of mechanism depending on the type of diodes we have chosen:

(i) Avalanche breakdown and (ii) Zener breakdown.

(i) **Avalanche Breakdown**: For the *pn* junction with doping of *p* and *n* less than 5×10^{17}/cc, thicker depletion width ≈ 1 μm will be there, then the breakdown mechanism is by the process of avalanche breakdown. In this mechanism,

    (a) The applied electric field in the depletion layer is sufficiently high ($\approx 10^6$ V/cm).

    (b) Minority carriers accelerates inside the depletion layer and collides with the semiconductor crystal atoms, 'knocking off', the electron of this atom. Thus, break its covalent band and creates electron–hole pairs in large number.

    (c) The pair of electron–hole is created in the midst of the high field; they quickly separate and attain high velocities as the depletion region is long enough to cause further pair generation through more collisions.

    (d) Cumulative process of multiplications of carriers at the breakdown voltage giving rise to an almost infinite current. This process is known as Avalanche breakdown at a voltage (V_A) with large current as shown in Fig. 2.10a. The diode gets 'burnt off' unless and until large series resistance is put with the *dc* source for controlling the current.

(ii) **Zener Breakdown**: Zener breakdown takes place in *pn* junction diode having very thin depletion region ≈ 0.1 μm and this is when both the sides are heavily doped ($\gg 5 \times 10^{17}$/cc). This breakdown can be explained by band theory as well as by atomic bond theory:

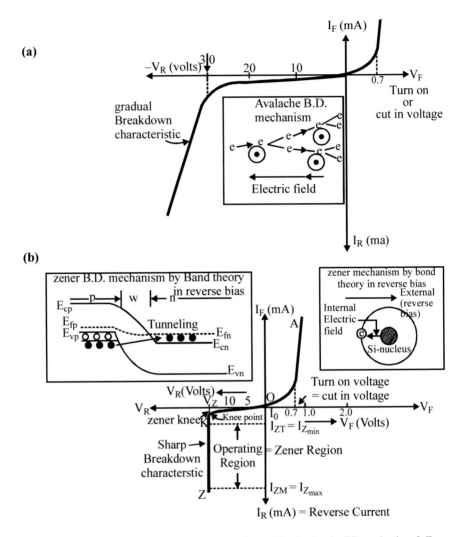

Fig. 2.10 **a** Avalanche breakdown I-V characteristics and the Avalanche BD mechanism. **b** Zener breakdown I-V characteristic and the Zener BD mechanism explained by band theory and covalent bond theory, with reverse bias

(a) Due to large doping and the reverse bias the level of conduction band of n side becomes lower than valance band of p side (see Fig. 2.10b) $E_{cn} < E_{vp}$. Therefore, at reverse bias, the electrons of p side (valance bond) jump (tunnels) to the conduction band of n side. This cross movement of electron is called tunnelling of electron, by which the current increases very much leading to Zener breakdown.

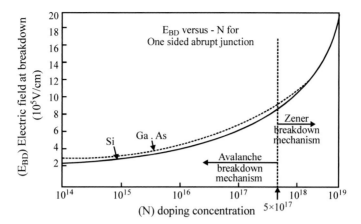

Fig. 2.11 Electric field at breakdown versus doping concentration in Si and GaAs for one-sided abrupt junction at 300 °K

(b) As per bond theory, we know that the electrons are bonded with the nucleus of the atom by the internal electric field (E_{int}) between them. Therefore, even at small reverse bias of units of volts, the electric field (E_{ext}) increases very much ($\geq 10^6$ V/cm) as the depletion layer is very thin ≈ 0.1 μm. This field E_{ext} being greater than E_{int}, the electrons are snatched away from the atom (i.e. the covalent bond is torn off), leading to multiplication of carriers and hence the breakdown.

Here, in Zener diodes, electric field is slightly higher than in Avalanche diode (see Fig. 2.11), but the free minority electrons do not have enough length to accelerate for knocking off the electrons of Si atoms, and therefore avalanche mechanism do not occur here.

Thus, we conclude that (a) in both the cases of breakdown (Avalanche and Zener), the covalent bond gets broken and carrier multiplication takes place increasing the current (b) multiplication of carriers in Avalanche by 'knocking off' of electron of Si atom by high-speed electron, while in Zener diodes, it is by snatching away of electron due to high external field. (c) In avalanche diodes, doping is $<5 \times 10^{17}$/cc while in Zener diodes, it is $>5 \times 10^{17}$/cc. (d) This causes depletion region thickness to be ≈ 1.0 μm and ≈ 0.1 μm, respectively. (e) In both the cases, diodes can burn unless the current is controlled by external resistance in series with the bias.

2.2.2 Comparison of Zener and Avalanche Breakdown Diode

S. no.	Parameters and properties	Zener breakdown diode	Avalanche breakdown diode
1.	Material used	Si	Si
2.	Depletion layer width(W) $$\sqrt{\frac{2\in V_0}{q}\left(\frac{1}{N_A}+\frac{1}{N_D}\right)}$$	Narrow; $W = \approx 0.1\,\mu m$ Very thin	Wide $W = \approx 1.0\,\mu m$ Thicker
3.	Doping densities N_D, N_A	$> 5 \times 10^{17}/cc$	$< 5 \times 10^{17}/cc$
4.	(a) V_Z (breakdown voltage) (b) Turn-on voltage (V_{BD}) $\left(\frac{qW^2}{2\varepsilon}\right)\left(\frac{N_A \cdot N_D}{N_A+N_D}\right)$	$V_Z \sim 6\,V$ or less $V_0 \approx 0.75\,V$	$V_Z > 6\,V$ $V_0 = 0.6\,V$
5.	Energy band and Fermi level at zero-level bias		
6.	Equivalent circuit and symbol	(In reverse bias)	(At reverse bias)
7.	The electron carrier increase is the result of high electric field E_{ext} causing	Breaking of atom-e-bond in thin depletion layer $E_{ext} > E_{in}$, hence snatching away of electrons. Here $E_{ext} > 10^6$ V/cm. As per band theory, cross-band tunnelling of electron takes place	Knocking off of 'e' of atom by high-speed minority 'e' in thick depletion layer. Here $E_{ext} \le 10^6$ V/cm
8.	Variation of breakdown voltage with temperature	The breakdown voltage decreases with increase in temperature	The breakdown voltage increases with increase in temperature
9.	V-I characteristics of Avalanche diode and Zener diode in forward and reverse bias	(a) Reverse bias V-I characteristic with the Zener breakdown is very sharp at V_Z with tunnelling current (b) In forward bias, majority current flows after cut-in	(a) Reverse bias V-I characteristic with avalanche breakdown is gradual near V_a as well as beyond

(continued)

(continued)

S. no.	Parameters and properties	Zener breakdown diode	Avalanche breakdown diode
		voltage $V_0 = 0.75$ V (called turn-ON voltage). Hence, no tunnelling	(b) In forward bias majority current flows after cut-in voltage $V_0 = 0.6$ V (called turn-ON voltage). Here, no tunnelling is there neither in forward nor in reverse bias
10.	Power dissipation at reverse bias (in general)	100 mW – 50 mW (As per area)	100 mW – 50 mW (As per area)
11.	Temperature coefficient of breakdown voltage (V_{BD}) and leakage current (I_S)	$\partial V_{BD}/\partial T = -\text{ve}$ $\partial I_S/\partial T = +\text{ve}$	$\partial V_{BD}/\partial T = +\text{Ve}$ $\partial I_S/\partial T = +\text{Ve}$
12.	Application	(a) Voltage Regulator (b) Device circuit protector in MOS circuit (c) Reference voltage	(a) Microwave signal generator as IMPATT/ TRAPATT diode (b) Temperature compensator to Zener diode in series in voltage regulator

2.2.3   Zener Diode Characteristics and Specification

The V-I characteristics of a Zener diode is shown in Fig. 2.10b. The following points are worth noting:

(i) Its characteristics are similar to an ordinary rectifier diode with the exception that it has a sharp breakdown voltage called Zener voltage V_Z.

(ii) It can be operated (see Fig. 2.10b) in any of the three regions, i.e. forward, leakage (OK), or breakdown (KZ). But usually it is operated in the breakdown region (KZ) for using it as voltage regulator.

Although a Zener diode can be operated in the forward region like an ordinary diode and can be used as for rectifier in reverse bias but not used due to its heavy cost.

(iii) The voltage is almost constant (V_Z) over the operating region, also called as Zener region (KZ in Fig. 2.10b).

(iv) Usually, the value of Zener voltage (V_Z) at a particular test current I_{ZT} ($=I_{Zmin}$) is specified in V-I characteristics shown in Fig. 2.10b.

(v) During operation, it will not burn as long as the external circuit limits the current flowing through it below the burnt-out value, i.e. I_{ZM} ($=I_{Zmax}$). Below this burnt-out value (I_{Zmax}) is the maximum rated Zener current.

(vi) Generally, manufacturers provide a separate data sheet for various Zener diodes depending upon their V-I characteristics. Some of the important ratings for a given typical Zener diode are as follows:

Specifications:

Zener diode (V_Z)	9 V $\pm$ 10%, i.e. valid from (8.1 V–9.9 V) range
Tolerance	10%
Effect of Temperature	Decrease is about 2 mV/°C rise in temperature
Power Ratings	(0.25–50) W
I_{Zmin}	7 mA
I_{Zmax}	48 mA (max)

Circuit symbol of Zener diode

2.2.4 Zener Diode as a Voltage Regulator

Zener diodes find wide commercial and industrial applications. The major application of a Zener diode in the electronic circuit is as a voltage regulator. It provides a constant voltage to the load from a source whose voltage may vary over sufficient range. Figure 2.12 shows the circuit arrangement.

This circuit is also called as Zener diode shunt (voltage) regulator because the Zener diode is connected in parallel or shunt with the load. The resistance R_S is connected to limit current in the circuit. The output, i.e. regulated voltage is obtained across the load resistor R_S. For the operation of the circuit, the input voltage ($V_S = V_{in}$) should be greater than Zener diode voltage (V_Z). Only then Zener diode operates in breakdown region.

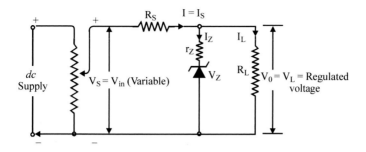

Fig. 2.12 Zener diode as voltage regulator

The input current is calculated by

$$V_S = I_S R_S + V_Z \tag{2.14}$$

$$I_S = \frac{V_S - V_Z}{R_S} \tag{2.14a}$$

Here

V_S = Input voltage (*dc* supply) unregulated (To be $> V_Z$)
$V_L = V_0 = V_Z$ = voltage across Zener = *dc* = Regulated voltage

The ideal Zener diode may be assumed as a constant voltage source of voltage V_Z. But in case of practical Zener diode, Zener resistance is taken into account. Due to this Zener resistance, there is a voltage drop ($I_Z r_Z$), also in addition to the voltage across the terminals of Zener diode; therefore, the load voltage may be given as

$$V_L = V_Z + I_Z r_Z \tag{2.15}$$

But the Zener resistance r_z being very very small ($<0.2\ \Omega$) may be neglected. Therefore, the load voltage is

$$V_L = V_Z \tag{2.16}$$

The current through the load resistance is

$$I_L = \frac{V_L}{R_L} = \frac{V_Z}{R_L} \tag{2.16a}$$

On applying the KCL, the input current will be

$$I_S = I_Z + I_L \tag{2.17}$$

$$I_Z = I_S - I_L \tag{2.17a}$$

2.2.4.1 Limitation of Zener Regulators

Even though the Zener diode provides a very simple means of voltage regulation, these regulators have the following disadvantages:

(i)   Since the output voltage of Zener regulator is a constant voltage (V_Z), therefore these voltage regulators cannot be made adjustable.

(ii) Large amount of power gets dissipated in the series resistance R_S and hence power wastage.

(iii) Corresponding to large changes in the Zener load current, there will be large changes in the Zener current leading to huge power wastage.

2.2.5 Effect of Temperatures on Zener Diodes

It is well known from previous sections that *pn* junction is temperature sensitive both in forward bias and in reverse bias. Since Zener diode is used only in reverse bias, the leakage current or minority carries will be temperature dependent and therefore the Zener voltage (V_Z) also becomes temperature dependent, because of drops across its series resistance (r_z). Therefore, normally Zener diodes' breakdown voltage have −ve voltage temperature coefficient (α). This characteristic α_z is usually included in the applicable Zener diode specification data sheet, where it is often stated in a percent in Zener voltage per degree centigrade (% C), or occasionally in mV/°C. The α_z can be as low as −0.09%/°C for low-voltage zeners, or as high as +0.11%/°C for high-voltage zeners. This is illustrated in Fig. 2.13. The temperature coefficient is used in predicting voltage temperature behaviour. Figure 2.13 shows that a negative to positive temperature coefficient 'transition'

Fig. 2.13 Temperature coefficient versus Zener break down voltage (V_Z)

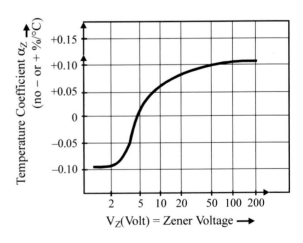

Fig. 2.14 Zener diode circuit

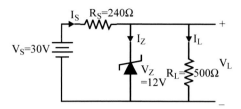

will occur in the vicinity of 5 V, and the positive α_z is there due to Avalanche breakdown which is generally independent of operating current (I_Z).

Zeners are specified for Zener voltage (V_Z) at ambient temperature (T_A) of 25 °C. The voltage change (ΔV_Z) may be calculated for its Zener voltage as

$$(\Delta V_Z) = \alpha_Z \times V_Z \times \Delta T_A / 100$$

Here, ΔT_A is a Zener diode temperature change.

Example 2.8 A Zener diode shunt regulator is shown in Fig. 2.14. Determine (i) the load voltage, (ii) voltage drop across R_S and (iii) current through the Zener diode.

Solution Given that

$$V_S = 30\,\text{V}, \quad V_Z = 12\,\text{V}$$
$$R_S = 240\,\Omega, \quad R_L = 500\,\Omega$$

$$V_S = I_S R_S + V_Z$$

(i)
$$I_S = \frac{V_S - V_Z}{R_S} = \frac{30 - 12}{240} = \frac{18}{240} = 0.075\,\text{A}$$

Voltage drop across $R_S = I_S R_S$
$$= 0.075 \times 240 = 18\,\text{V}$$

(ii)
$$I_L = \frac{V_L}{R_L} = \frac{12}{500} = 0.024\,\text{A}$$

$$\therefore \text{Current through the Zener diode}(\text{IZ}) = I_S - I_L$$
$$= 0.075 - 0.024$$
$$= 0.051\,\text{A}$$

Example 2.9 For the circuit shown in Fig. 2.15, find

(i)   the output voltage,
(ii) the voltage drop across the series resistance and
(iii) the current through Zener diode.

Fig. 2.15 Zener diode circuit

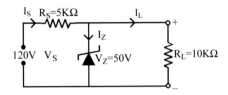

Solution Regarding the given figure,

(a) Output voltage $= V_L = V_Z = 50$ V.
(b) Voltage drop across the series resistance, R_S is

$$= I_S R_S = V_S - V_Z = 120\text{–}50 = 70 \text{ V.}$$

(c) Load current $(I_L) = \frac{V_L}{R_L} = \frac{50}{10\,k\Omega} = 0.005$ Amp

$$I_S = \frac{70 \text{ V}}{5 \text{ k}\Omega} = 0.014 \text{ Amp}$$
$$I_S = I_L + I_Z$$
$$\therefore \quad I_Z = I_L - I_Z$$
$$= 0.014 - 0.005$$
$$= 0.009 \text{ Amp}$$
$$= 9 \text{ mA}$$

Hence, the current through Zener diode is 9 mA.

2.3   Rectifiers

The electrical power is generated, distributed and transmitted as *ac* for economical reason. But most of the electronic circuits need *dc* voltage power supply for their operation.

Therefore, a circuit that converts *dc* voltage of main supply into *dc* voltage is called power supply. The block diagram of a power supply is shown in Fig. 2.16. Basically, at the input of the power supply, a stepdown transformer is used to step down the voltage as per need and is known as power transformer. It is followed by a diode circuit called rectifier which gives pulsating *dc* output, which is then fed to the filter circuit and finally to a regulator so as to obtain regulated *dc* at the output. Here, Zener diode acts as regulator.

Usually, the rectifier is the heart of power supply. Rectifiers are classified as half-wave or full-wave rectifier depending upon the type of output.

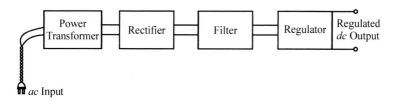

Fig. 2.16 Block diagram of a power supply

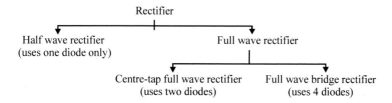

2.3.1 Half-Wave Rectifier

It converts *ac* voltage into a pulsating *dc* voltage and uses only one half of the applied *ac* voltage. Figure 2.17 shows the basic circuit and waveforms of a half-wave rectifier.

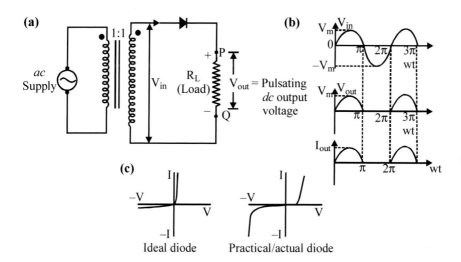

Fig. 2.17 a Basic circuit of the half-wave rectifier and **b** input and output of half-wave rectifier. **c** Ideal and actual diode characteristic

Let V_{in} be the alternating voltage and given by the equation

$$V_{in} = V_m \sin \omega t; \quad V_m \gg V_r$$

where V_r is the cut-in voltage of the diode. During the positive half cycle of the input signal, the anode of the diode becomes more positive with respect to the cathode making it forward biased and hence, diode D conducts with voltage across as zero, considering it an ideal diode as shown in Fig. 2.17c. So the whole input voltage will appear across the load resistance, R_L, with P point +ve and Q points −ve.

During negative half cycle of the input signal, the diode D does not conduct, as it gets reverse biased. For an ideal diode, the reverse-bias impedance offered by the diode is infinity with no current (Fig. 2.17c). So, the whole input voltage appears across diode D. Hence, the voltage drop across R_L is zero.

(a) **Ripple factor** (Γ): The ratio of RMS value of *ac* component to the *dc* component in the output is called ripple factor (Γ).

$$\Gamma = \frac{\text{RMS value of } ac \text{ component}}{dc \text{ value of component}} = \frac{V_{r,rms}}{V_{dc}}$$

where $V_{r,rms} = \sqrt{V_{rms}^2 - V_{dc}^2}$

$$\Gamma = \sqrt{\left(\frac{V_{rms}}{V_{dc}}\right)^2 - 1} \tag{2.9}$$

$$V_{av} = \left[\int_0^\pi V_m \sin wt.d(wt) + \int_0^{2\pi} 0.d(wt)\right]$$
$$= \frac{V_m}{2\pi}[-\cos wt]_0^\pi = \frac{V_m}{\pi} \tag{2.10}$$

This being the average of the +ve voltage, is nothing but the *dc* voltage $V_{dc} = V_{av} = V_m/\pi$.

Therefore,

$$I_{dc} = \frac{V_{dc}}{R_L} = \frac{V_m}{\pi R_L} = \frac{I_m}{\pi} \tag{2.11}$$

Here I_m is the peak current corresponding to maximum voltage V_m.

If the values of diode forward resistance (r_f) are taken into account, then

$$I_{dc} = \frac{V_m}{\pi(R_L + r_f)} \qquad (2.11a)$$

The RMS voltage at the load resistance can be calculated as

$$V_{rms} = \sqrt{\frac{1}{2\pi}\int_0^\pi (V_m \sin(\omega t))^2 d(\omega t)}$$

$$= \sqrt{\frac{V_m^2}{4\pi}[1 - \cos(2\omega t)]_0^\pi d(\omega t)} = \frac{V_m}{2} \qquad (2.12)$$

$$V_{rms} = \frac{V_m}{2}$$

Therefore, ripple factor

$$(\Gamma) = \sqrt{\frac{\left(\frac{V_m}{2}\right)^2}{\left(\frac{V_m}{\pi}\right)^2} - 1} = \sqrt{\left(\frac{\pi}{2}\right)^2 - 1} = 1.21$$

From this expression, it is understood that the average amplitude of *ac* signal, present in the output, is 121% of the *dc* voltage. So the half-wave rectifier is not practically useful in converting *ac* into *dc*.

(b) **Efficiency (η):** The ratio of *dc* output power to *ac* input power is called rectifier efficiency.

$$\eta = \frac{dc \text{ output power}}{ac \text{ input power}} = \frac{P_{dc}}{P_{ac}}$$

$$= \frac{\frac{(V_{ac})^2}{R_L}}{(V_{rms})^2/R_L} = \frac{4}{\pi^2} = 0.406 = 40.6\%$$

The maximum efficiency of a half-wave rectifier is only 40.6% as the half wave is only used.

(c) **Peak inverse voltage (*PIV*):** It is defined as the maximum reverse voltage that a diode has to withstand in the circuit.

In other words, the peak inverse voltage across a diode is the peak of the negative half cycle of input voltage. For half-wave rectifier, PIV is V_m.

(d) **Transformer Utilization factor (TUF):** For any power supply design, the transformer rating should be determined. It can be done with a knowledge of the *dc* power delivered to the load and *ac* rating of the secondary of the transformer

$$TUF = \frac{dc \text{ power delivered to the load}}{ac \text{ rating of the transformer secondary}}$$

$$= \frac{P_{dc}}{P_{ac\,rated}} = \frac{P_{ac}}{P_{rms}}$$

For half-wave rectifying circuit, the rated voltage of the transformer secondary is $V_m/\sqrt{2}$, but the actual RMS current flowing through the winding is only $\frac{I_m}{2}$ and not $\frac{I_m}{\sqrt{2}}$

$$TUF = \frac{(I_{dc})^2 R_L}{V_{rms} I_{rms}} = \frac{\left(\frac{I_m}{\pi}\right)^2 R_L}{\frac{V_m}{\sqrt{2}} \cdot \frac{I_m}{2}} \qquad (Also \quad V_m = I_m \cdot R_L)$$

$$= \frac{2\sqrt{2}}{\pi^2} = 0.287$$

Thus, the TUF for a half-wave rectifier is 0.287.

(e) **Form factor**:

$$Form\,factor = \frac{RMS\ value}{Average\ value} = \frac{V_m/2}{V_m/\pi} = 1.57$$

(f) **Peak factor**:

$$Peak\,factor = \frac{Peak\ value}{RMS\ value} = \frac{V_m}{V_m/2} = 2$$

2.3.2   Full-Wave Rectifier

It converts an *ac* voltage into a pulsating *dc* voltage using both half cycles of the applied *ac* voltage. There are two types of full-wave rectifier

(i) Centre-tap full-wave rectifier.
(ii) Bridge-type full-wave rectifier.

2.3.2.1 Full-Wave Centre-Tap Rectifier

This full-wave rectifier uses two diodes of which one diode conducts during positive half cycle, while the other diode conducts during negative half cycle of the applied *ac* voltage of other half winding of the secondary of transformer. Figure 2.18 shows the basic circuit and waveform of full-wave centre-tapped rectifier. The secondary winding of the transformer has a centre-tap point (C) normally grounded, with equal number of windings across it.

During positive half of the input signal, anode of diode D_1 becomes positive and at the same time, anode of the diode D_2 becomes negative. Hence, diode D_1 conducts and diode D_2 does not conduct. The load current flows through diode D_1 and the voltage drop across R_L will be equal to the input voltage, with P point +ve and Q as −ve.

During negative half of the input signal, the anode of diode D_1 becomes negative and the anode of diode D_2 becomes positive. Hence, diode D_1 does not conduct and the diode D_2 conducts. The load current flows through the diode D_2 and the voltage drop across R_L will be equal to the input voltage, in the same direction again with P point +ve and Q as −ve.

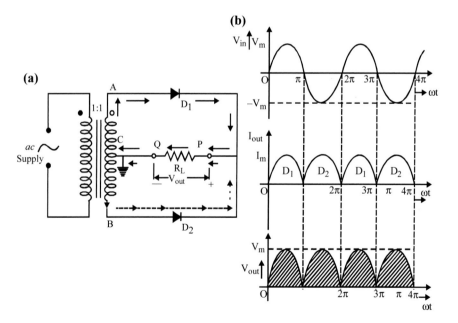

Fig. 2.18 a Centre-tap full-wave rectifier, **b** input and output waveforms of full-wave rectifier

(a) **Ripple factor** $(\Gamma) = \sqrt{\frac{V_{rms}^2}{V_{dc}^2} - 1}$

The average voltage or *dc* voltage available across the load resistance is

$$
\begin{aligned}
V_{dc} &= \frac{1}{2\pi}\left[\int_0^\pi V_m \sin \omega t\, d(\omega t) + \int_\pi^{2\pi} V_m \sin \omega t\, d(\omega t)\right] \\
&= \frac{1}{\pi}\left[\int_0^\pi V_m \sin \omega t\, d(\omega t)\right] \\
&= \frac{V_m}{\pi}\left[-\cos \omega t\right]_0^\pi = \frac{2V_m}{\pi} \\
I_{dc} &= \frac{V_{dc}}{R_L} = \frac{2I_m}{\pi} = \frac{2V_m}{\pi R_L}
\end{aligned}
$$

The RMS value of the voltage at the load resistance (R_L) is

$$
V_{rms} = \sqrt{\frac{1}{\pi}\int_0^\pi V_m^2 \sin^2 \omega t\, d(\omega t)} = \frac{V_m}{\sqrt{2}}
$$

$$
\therefore \qquad \Gamma = \sqrt{\frac{\left(\frac{V_m}{\sqrt{2}}\right)^2}{\left(\frac{2V_m}{\pi}\right)^2} - 1} = \sqrt{\frac{\pi^2}{8} - 1} = 0.482
$$

Hence, it is clear that the average amplitude of *ac* present in the output is 48.2% of the *dc* voltage.

(b) **Efficiency**

$$
(\eta) = \frac{dc \text{ output power}}{ac \text{ input power}} = \frac{P_{dc}}{P_{ac}}
$$

$$
\frac{V_{dc}^2/R_L}{(V_{rms})^2/R_L} = \frac{\left(\frac{2V_m}{\pi}\right)^2}{\left(\frac{V_m}{\sqrt{2}}\right)^2} = \frac{8}{\pi^2} = 0.812 = 81.2\%
$$

Therefore, the maximum efficiency of a full-wave rectifier is 81.2%.

(c) **Transformer Utilization Factor (TUF)**

The average TUF of a full-wave rectifying circuit is determined by considering the primary and secondary winding separately:

$$\text{TUF of secondary winding} = \frac{V_{dc}I_{dc}}{V_{rms}I_{rms}} = \frac{(I_{dc})^2 R_L}{\frac{V_{rms}^2}{R_L}} = \frac{\frac{4}{\pi^2}\frac{V_m^2}{R_L}}{\frac{V_m^2}{2R_L}} = 0.812$$

$$\text{TUF of primary winding} = 2 \times \text{TUF of half wave rectifier}$$
$$= 2 \times 0.287 = 0.574$$

$$\text{Average TUF} = \frac{\text{TUF (Primary)} + \text{TUF (Secondary)}}{2}$$
$$= \frac{0.812 + 0.574}{2} = 0.693$$

(d) **Form factor**

$$FF = \frac{\text{RMS value}}{\text{Average value}} = \frac{\frac{V_m}{\sqrt{2}}}{\frac{2V_m}{\pi}} = \frac{\pi}{2\sqrt{2}} = 1.11$$

(e) **Peak factor**

$$PF = \frac{\text{Peak value}}{\text{RMS value}} = \frac{V_m}{\frac{V_m}{\sqrt{2}}} = \sqrt{2}$$

(f) **PIV**: The Peak Inverse Voltage (PIV) for full-wave rectifier for centre tapping is $2V_m$ because the entire secondary voltage appears across the nonconducting diode.

2.3.2.2 Full-Wave Bridge Rectifier

The need for a costly centre-tapped transformer in a full-wave rectifier is eliminated in the bridge rectifier as the diodes are very cheap. As shown in Fig. 2.19, the bridge rectifier using four diodes is connected to form a bridge. The *ac* input voltage is applied to diagonally opposite ends of the bridge. The load resistance (R_L) is connected between the other two ends of the bridge with directions of the diode terminals as shown in Fig. 2.19.

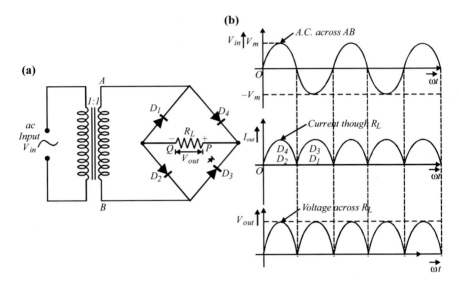

Fig. 2.19 a Bridge rectifier, **b** input and output waveforms of bridge rectifier

For the positive half cycle of the *ac* input voltage, the diodes D_4 and D_2 conduct, whereas diodes D_3 and D_1 do not conduct. The conducting diodes will be in series through the load resistance (R_L). So, the load current flows through R_L, from *P* to *Q*.

During the negative half cycle of the *ac* input voltage, the diodes D_3 and D_1 conduct, whereas diodes D_4 and D_2 do not conduct. The current conducting diodes D_2 and D_4 will be in series through the load resistance (R_L) and in the same direction from *P* to *Q* as in the previous half cycle. Thus, a bidirectional wave is converted into a unidirectional one for load (R_L).

The average values of output voltage, load current and RMS value of output voltage for bridge rectifier are exactly same as for a centre-tapped full-wave rectifier, using the ideal diode.

$$V_{dc} = \frac{2V_m}{\pi}; \quad I_{dc} = \frac{V_{dc}}{R_L} = \frac{2V_m}{\pi R_L} = \frac{2I_m}{\pi}$$

$$V_{rms} = \frac{V_m}{\sqrt{2}}, I_{rms} = \frac{I_m}{\sqrt{2}}$$

(a) Maximum efficiency of a bridge rectifier = maximum efficiency of a centre-tapped rectifier = 81.2%.
(b) Ripple factor (Γ) = 0.482 = same as centre-tapped rectifier.
(c) PIV = V_m, as a result, diode life is longer than in centre-tap rectifier.

2.3.2.3 Advantage of Bridge Rectifier Over Centre-Tapped Full-Wave Rectifier

The ripple factor (Γ) and efficiency (η) of bridge rectifier are exactly same as for centre-tapped full-wave rectifier but has a number of advantages:

(a) Need for having a centre taping with double the secondary winding is eliminated.
(b) **Reduced need of PIV of diodes**: The PIV across either of the nonconducting diodes is equal to the peak value of the transformer secondary voltage V_m as compared to bridge rectifier is only half of that for a centre-tapped full-wave rectifier. This is a great advantage, which offsets the disadvantage of using extra two diodes in a bridge rectifier.
(c) **Higher TUF**: One more reason that the bridge rectifier is better and popular over centre-tapped rectifier is from TUF point of view. Since TUF of bridge rectifier is 0.812, which considerably high over TUF of centre-tapped rectifier (0.693). It is because the current flowing in the transformer secondary for bridge rectifier is purely alternating, so the TUF increases to 0.812.
(d) **Floating Point Output is there**: The bridge rectifiers are used in applications allowing floating output terminals, i.e. no output terminal is grounded.

2.3.2.4 Comparison of Rectifiers

S. no.	Particulars	Half-wave	Full-wave	
			Centre tap	Bridge
1.	No. of diodes used	1	2	4
2.	Transformer necessary	No	Yes, with double winding in secondary	No
3.	V_{dc} (ideal case)	V_m/π	$2V_m/\pi$	$2V_m/\pi$
4.	dc current	I_m/π	$2I_m/\pi$	$2I_m/\pi$
5.	Ripple factor	1.21	0.482	0.482
6.	Maximum efficiency	40.6%	81.2%	81.2%
7.	Peak inverse Voltage (PIV) requirement of diodes	V_m	$2V_m$	V_m
8.	Transform utilization factor	0.287	0.693	0.812
9.	Form factor	1.57	1.11	1.11
10.	Peak factor	2	$\sqrt{2}$	$\sqrt{2}$
11.	Output ripple frequency	f	$2f$	$2f$

Example 2.10 A half-wave rectifier has a resistive load $R_L = 1000\,\Omega$, with an *ac* voltage 325 *V* (peak value) and the diode has a forward resistance of $100\,\Omega$. Calculate (a) peak average and RMS value of current, (b) *dc* output power, (c) *ac* input power and (d) efficiency of the rectifier.

Solution

(a) Peak value of current,

$$I_m = \frac{V_m}{r_f + R_L} = \frac{325}{100 + 1000} = 295.45 \text{ mA}$$

$$\text{Average current } I_{dc} = \frac{I_m}{\pi} = \frac{295.45}{3.14} = 94.046 \text{ mA}$$

$$\text{RMS current } I_{rms} = \frac{I_m}{2} = 147.725 \text{ mA}$$

(b) *dc* power output $= I_{dc}^2 \times R_L$

$$= \left(94.046 \times 10^{-3}\right)^2 \times 1000$$

$$= 8.845 \text{ W}$$

(c) *ac* input power $= I_{rms}^2 (r_f + R_L)$

$$= (147.725 \times 10^{-3})^2 (100 + 1000)$$

$$= 24 \text{ W}$$

(d) Efficiency$(\eta) = \dfrac{P_{dc}}{P_{ac}} = \dfrac{8.845}{24} \times 100\%$

$$= 36.85\%$$

Example 2.11 A half-wave rectifier is shown in Fig. 2.20. Find maximum and average values of power delivered to the load. Assume $R_L = 200\,\Omega\ (N_1 : N_2 = 2 : 1)$.

Solution $N_1 : N_2 = 2 : 1$

$$V_{in} = V_1 = 230 \text{ V,}$$

Fig. 2.20 Half wave rectifier circuit

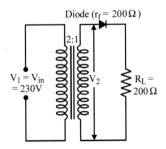

As we know,

$$V_2 = \frac{V_m}{\sqrt{2}} = \frac{N_2}{N_1} V_1 = \frac{1}{2} \times 230$$

$$\therefore \quad \frac{V_m}{\sqrt{2}} = 115$$

$$V_m = 115\sqrt{2} = 162.6\,\text{V}$$

$$I_m = \frac{V_m}{R_L} = \frac{162.6}{200} = 0.813\,\text{Amp}$$

Maximum power delivered to the load

$$P_m = I_m^2 R_L = (0.813)^2 \times 200$$
$$= 132.3\,\text{W}$$
$$V_{dc} = \frac{V_m}{\pi} = \frac{162.6}{\pi} = 51.7\,\text{V}$$
$$I_{dc} = \frac{V_{dc}}{R_L} = 0.26\,\text{Amp}$$
$$P_{average} = I_{dc}^2 \times R_L$$
$$= (0.26)^2 \times 200$$
$$= 13.52\,\text{W}$$

Example 2.12 A diode with internal resistance $r_f = 20\,\Omega$ in half-wave rectifier, is applied $V = 50 \sin \omega t$ and $R_L = 800\,\Omega$. Calculate (a) I_m, I_{dc} and I_{rms}, (b) *ac* power input and *dc* power output, (c) *dc* output voltages and (d) efficiency of rectifier.

Solution

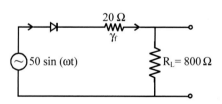

$$V = 50 \sin \omega t$$
$$V_m = 50\,\text{volt}$$

(a)
$$I_m = \frac{V_m}{r_f + R_L} = \frac{50}{20 + 800} = 0.061 \text{ A} = 61 \text{ mA}$$

$$I_{dc} = \frac{I_m}{\pi} = \frac{61}{\pi} = 19.4 \text{ mA}$$

$$I_{rms} = \frac{I_m}{2} = \frac{61}{2} = 30.5 \text{ mA}$$

(b)
$$ac \text{ power input} = I_{rms}^2 (r_f + R_L)$$
$$= (30.5 \times 10^{-3})^2 (20 + 800)$$
$$= 0.763 \text{ W}$$

$$d.c. \text{ power output} = (I_{dc})^2 R_L$$
$$= (19.4 \times 10^{-3})^2 \times 800$$
$$= 0.301 \text{ W}$$

(c)
$$d.c. \text{ output voltage} = I_{dc} R_L = 19.4 \text{ mA} \times 800 \,\Omega$$
$$= 15.2 \text{ V}$$

(d)
$$\text{Efficiency} = \frac{P_{dc}}{P_{ac}} \times 100\% = \frac{0.301}{0.763} \times 100\% = 39.5\%$$

Example 2.13 A full-wave rectifier delivers 50 W to a load of 200 Ω. If the ripple factor is 1%, calculate the *ac* ripple voltage across the load.

Solution

$$P_{dc} = \frac{V_{dc}^2}{R_L}$$

$$V_{dc} = \sqrt{P_{dc} \times R_L} = \sqrt{50 \times 200} = 100 \text{ V}$$

$$\Gamma = \frac{V_{ac}}{V_{dc}} = 1\% = 0.01$$

$$V_{ac} = 0.01 \times V_{dc} = 1 \text{ V}$$

Therefore, the *ac* ripple voltage across the load is $V_{ac} = 1$ V.

Example 2.14 A 230 V, 50 Hz voltage is applied to the primary of a 4: 1 stepdown transformer used in a bridge rectifier having a load resistance of 1000 Ω. Assuming the diode to be ideal, calculate (a) *dc* output voltage, (b) *dc* power delivered to the load, (c) PIV and (d) output ripple frequency.

Solution

(a) RMS value of transformer secondary voltage

$$V_{rms} = \frac{230}{4} = 57.5 \text{ V}$$

The maximum value of secondary voltage

$$V_m = \sqrt{2}V_{rms} = \sqrt{2} \times 57.5 = 81.3 \text{ V}$$

Therefore, *dc* output voltage,

$$V_{dc} = \frac{2V_m}{\pi} = \frac{2 \times 81.3}{\pi} = 51.8 \text{ V} \approx 52 \text{ V}$$

(b) *dc* power delivered to the load,

$$P_{dc} = \frac{V_{dc}^2}{R_L} = \frac{52^2}{1000} = 2.704 \text{ W}$$

(c) PIV $= V_m = 81.3$ V
(d) Output ripple frequency $= 2f = 2 \times 50 = 100$ Hz.

2.4   Filters

The output of a rectifier contains *dc* component as well as *ac* component. The filters are used to minimize the undesirable *ac*, i.e. ripple with the *dc* component of the output as in Fig. 2.21.

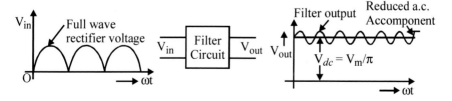

Fig. 2.21 Filter circuit

Some important types of filters are as follows:

(i)   Inductor filter
(ii)   Capacitor filter
(iii) LC or L-section filter
(iv)   CLC or π-type filter.

2.4.1 Inductor Filter

It is also called as series inductor filter. Figure 2.22a shows the inductor filter. An inductor (choke) is just connected in series with the load. The inductors have the inherent property to oppose the change of current. This property of the inductor is utilized here to suppress the *ac* component (ripples) from the output of a rectifier.

The reactance $(X_L = 2\pi f l)$ of the inductor is large for high frequencies and offers more opposition to them but it allows the *dc* component of the rectifier to flow to the load at output. Thus, it smooths out the rectifier output from V_{01} to V_0, as shown in Fig. 2.22b.

(a)

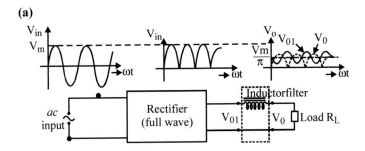

Fig. 2.22a a Inductor filter

Fig. 2.22b Rectifier output

(b)

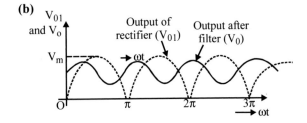

To analyse the filter for a full wave, Fourier series can be proved to be as

$$V_0 = \frac{2V_m}{\pi} - \frac{4V_m}{\pi}\left[\frac{1}{3}\cos 2\omega t + \frac{1}{15}\cos 4\omega t + \frac{1}{35}\cos 6\omega t + \cdots\right]$$

Here, the *dc* component is $\frac{2V_m}{\pi}$ assuming the third and higher terms contribute little, the output voltage will be having an *ac* ripple component of frequency 2ω

$$V_0 = \frac{2V_m}{\pi} - \frac{4V_m}{3\pi}\cos 2\omega t \tag{2.13}$$

The diode, choke and transformer resistances can be neglected since they are very small as compared with R_L. Therefore, the *dc* component of current $I_m = \frac{V_m}{R_L}$. The impedance in series of L and R_L at 2ω frequency is

$$Z = \sqrt{(R_L)^2 + (2\omega L)^2} = \sqrt{R_L^2 + 4\omega^2 L^2}$$

Therefore, for the maximum i.e. peak *ac* component

$$I_m = \frac{V_m}{Z} = \frac{V_m}{\sqrt{R_L^2 + 4\omega^2 L^2}} \text{ at } \phi = \tan^{-1}\left(\frac{2\omega L}{R_L}\right) \tag{2.14}$$

So, the resulting current i using Eqs. 2.13 and 2.14 is given by

$$i = \frac{2V_m}{\pi R_L} - \frac{4V_m}{3\pi}\frac{\cos(2\omega t - \phi)}{\sqrt{R_L^2 + 4\omega^2 L^2}} = (i_{dc} - i_{ac}) = i_{dc} - I_o \cdot \cos(2\omega t - \phi)$$

The ripple factor (Γ) is defined as the ratio of the RMS value of the ripple current to the *dc* current. The peak value of ripple current at frequency 2ω is $4V_m/\left(3\pi \cdot \sqrt{R_L^2 + 4\omega^2 L^2}\right)$; the RMS value of it will be obtained by dividing this by $\sqrt{2}$. Therefore,

$$\Gamma = \frac{\dfrac{4V_m}{3\sqrt{2}\pi\sqrt{R_L^2 + 4\omega^2 L^2}}}{\dfrac{2V_m}{\pi R_L}} = \frac{2}{3\sqrt{2}}\frac{1}{\sqrt{1 + \dfrac{4\omega^2 L^2}{R_L^2}}} \tag{2.15}$$

Case 1: $\frac{4\omega^2 L^2}{R_L^2} \gg 1$, then a simplified expression for ripple factor is

$$\Gamma = \frac{R_L}{3\sqrt{2}\omega L}$$

> Therefore, in general, ripple factor is proportional to the load and inversely to the inductance of inductor filter.

Case 2: When the load resistance is infinity, i.e. the output is an open circuit, then the ripple factor is

$$\Gamma = \frac{2}{3\sqrt{2}} = 0.471$$

This value is slightly less than the ripple factor of full-wave rectifier ($\Gamma = 0.482$). The difference being attributable to the omission of higher harmonics as mentioned. It is clear that the series inductor filter should only be used where R_L is consistently small.

Example 2.15 Calculate the value of inductances to be used in the inductor filter connected to a full-wave rectifier operating at 50 Hz to provide a *dc* output with 5% ripple for a $100\,\Omega$ load.

Solution The ripple factor for inductor filter is

$$\Gamma = \frac{R_L}{3\sqrt{2}\omega L}$$

$$0.05 = \frac{100}{3\sqrt{2}(2\pi \times 50 \times L)}$$

$$\Rightarrow \quad L = \frac{1}{6\sqrt{2} \times 0.05\pi} = 0.7508\,\text{H}$$

2.4.2 Capacitor Filter

It is also called as shunt capacitor filter. Figure 2.23a shows the capacitor filter. The capacitor is just connected in parallel (shunt) with the load. The property of a capacitor in series is that it allows *ac* component and blocks the *dc* compound. Thus, the operation of a shunt capacitor filter here is to short and bypass the ripple to ground, leaving the purer *dc* to appear at the output.

During positive half cycle, the capacitor charges up to the peak value of the transformer secondary voltage V_m and will try to maintain this value (by supplying its charge current) when the full-wave input drops to zero. This requires large capacitor of 1000 μF or so.

Thus, this capacitor discharge through R_L slowly until the transformer secondary voltage again increases to a value greater than the capacitor voltage. The diode conducts for a period which depends on the capacitor voltage (equal to the load voltage). The diode will conduct when the transformer secondary voltage becomes

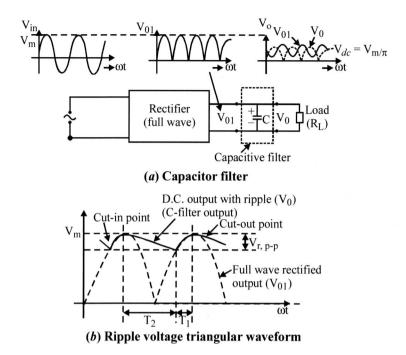

Fig. 2.23 a Capacitor filter, **b** ripple voltage triangular waveform

more than the 'cut-in' voltage of the diode. The diode stops conducting when the transformer voltage becomes less than the diode voltage. This is called cut-out voltage.

Figure 2.23b with slight approximation, the ripple voltage waveform can be assumed as triangular. From the cut-in point to the cut-out whatever charge the capacitor acquires is equal to the charge the capacitor has lost during the period of nonconductor, i.e. from cut-out point to the next cut-in point.

$$\text{The charge it has acquired} = V_{r,p-p} \times C$$
$$\text{The charge it has lost} = I_{dc} \times T_2$$
$$\text{Therefore,} \qquad V_{r,p-p} \times C = I_{dc} \times T_2$$

If the capacitor value is fairly large, or the value of load resistance (R_L) is very large, then it can be assumed that the time T_2 is equal to half the periodic time of the waveform.

$$\text{i.e.,} \qquad T_2 = \frac{T}{2} = \frac{1}{2f}, \text{ then } V_{r,p-p} = \frac{I_{dc}}{2fC} \qquad (2.16)$$

On considering the above assumptions, the ripple waveform can be taken to be triangular of height $V_{rp \cdot p}$ and therefore the RMS value of the ripple can be proved to be

$$V_{r,rms} = \frac{V_{r,p-p}}{2\sqrt{3}} \tag{2.17}$$

Therefore, from the above two equation no. Eqs. (2.16) and (2.17), we have the expression of $V_{r,rms}$ as:

$$V_{r,rms} = \frac{I_{dc}}{4\sqrt{3}fC} = \frac{V_{dc}}{4\sqrt{3}fCR_L}$$

$$\text{Therefore, ripple factor } \Gamma = \frac{V_{r,rms}}{V_{dc}} = \frac{1}{4\sqrt{3}fCR_L} \tag{2.18}$$

The ripple is inversely proportional to the load resistance and the capacitance in a capacitance filter.

The ripple may be decreased by increasing C or R_L (or both), resulting in increase in dc output voltage. If $f = 50$ Hz, C in μF and R_L in Ω, then

$$\Gamma = \frac{2890}{C\,R_L} \tag{2.19}$$

Example 2.16 Calculate the minimum value of capacitance in use in a capacitor filter connected to a full-wave rectifier operating at a standard aircraft power frequency of 100 Hz, if the maximum ripple factor allowed is 10% for a load of 500 Ω.

Solution

$$\Gamma = \frac{1}{4\sqrt{3}fC\,R_L}$$

$$0.01 = \frac{1}{4\sqrt{3} \times 100 \times C \times 500} = \frac{2.844 \times 10^{-6}}{C}$$

$$C = \frac{2.844 \times 10^{-6}}{0.01} = 284.4 \times 10^{-6} = 284.4\,\mu F$$

i.e. $C \geq 284.4\,\mu F$

2.4.3 LC Filter

We know that the ripple factor (Γ) is directly proportional to the load resistance (R_L) in the inductor filter and inversely proportional to the load resistance (R_L) in the capacitor filter. Therefore, if these two filters are combined as LC filter or

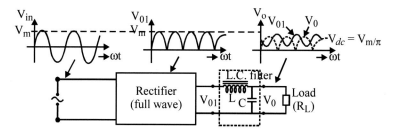

Fig. 2.24 LC filter

L-section filter as shown in Fig. 2.24, the ripple factor (Γ) will be independent of R_L. This advantage is not there in CLC π filter.

If the inductor value is increased, then it will increase the time of conduction. At some critical value of inductance, one diode D_1 or D_2 in full-wave centre-tapped rectifier will always be conducting.

From Fourier series, the output voltage can be expressed as

$$V_0 = \frac{2V_m}{\pi} - \frac{4V_m}{3\pi} \cos 2\omega t$$

The *dc* output voltage $V_{dc} = \frac{2V_m}{\pi}$ and peak value of ripple $ac = \frac{4V_m}{3\pi}$. The RMS ripple voltage will be $\frac{4V_m}{3\pi\sqrt{2}} = V_{rms}$.

Therefore,

$$I_{rms} = \frac{4V_m}{3\pi\sqrt{2}} \cdot \frac{1}{X_L} = \frac{\sqrt{2}}{3} \frac{V_{dc}}{X_L}$$

This current will flow through X_C, which bypasses the ripple voltage and does not allow it to reach the output

Therefore, $V_{r,rms} = I_{rms} \cdot X_c$

$$= \frac{\sqrt{2}}{3} \frac{V_{dc}}{X_L} \cdot X_c = \frac{\sqrt{2}}{3} V_{dc} \frac{X_C}{X_L} \qquad (2.20)$$

Then the ripple factor (Γ) will be

$$\Gamma = \frac{V_{r,rms}}{V_{dc}} = \frac{\sqrt{2}}{3} \frac{X_C}{X_L} = \frac{\sqrt{2}}{3} \frac{1}{\omega L} \cdot \frac{1}{\omega C} \qquad (2.21)$$

$$\Gamma = \frac{\sqrt{2}}{3\omega^2 LC} = \frac{\sqrt{2}}{3 \times 4\pi^2 f^2 LC} \qquad (2.22)$$

If $f = 50$ Hz, here C is in μF range and L is in Henry range, then

$$\text{ripple factor} \ (\Gamma) = \frac{1.194}{LC} \tag{2.23}$$

i.e. inversely proportional to the capacitance and inductance (in LC filter).

Advantages

(a) Ripple factor is very low (<1%).
(b) Ripple factor is independent of load resistance.
(c) Such filters are suitable for both heavy and low loads.

Disadvantages

(a) It cannot be used for half-wave rectifiers.
(b) The amplitude of output voltage is less.
(c) Inductances are more bulky and occupies more space.

Example 2.17 Calculate the ripple factor, when a full-wave rectifier is used as LC filter with $L = 20$ H and $C = 15$ μF. Assume the applied voltage is $V = 300 \sin (2\pi \times 50)t$.

Solution For LC filter, if C is in μF range and L is in Henry range, then ripple factor

$$(\Gamma) = \frac{1.194}{LC} = \frac{1.194}{20 \times 15}$$
$$= 0.00398$$
$$= 0.398\%$$

2.4.4 Multiple LC Filters

To get the better output from LC filters, two or more L-section filters are used, i.e. two or more L filters. It is shown in Fig. 2.25.

The ripple factor, for two LC filters (using Eq. 2.21), will be

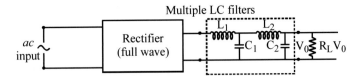

Fig. 2.25 Multiple LC filters

$$\Gamma = \frac{\sqrt{2}}{3} \cdot \frac{X_{C_2}}{X_{L_2}} \cdot \frac{X_{C_1}}{X_{L_1}} \tag{2.24}$$

2.4.5 CLC or π-Type Filter

Figure 2.26 shows the *CLC* or π-type filter which basically consists of a capacitor filter followed by an *LC* section.

This π-type filter offers a fairly smooth output and is characterized by a highly peaked diode currents and poor regulation.

The Fourier series of CLC filter is given by

$$V_0 = V_{dc} - \frac{2V_r}{\pi}\left(\sin 2\omega t - \frac{\sin 4\omega t}{2} + \frac{\sin 6\omega t}{3} - \cdots\right)$$

As we know that from capacitor filter analysis,

$$V_r = \frac{I_{dc}}{2fC}$$

the RMS of second harmonic voltage is

$$V_2' = \frac{2V_r}{\pi\sqrt{2}} = \frac{2I_{dc}}{2\pi fC_1\sqrt{2}} = \sqrt{2}I_{dc}X_{c_1}$$

where X_{c_1} is the reactance of C_1, at the second harmonic frequency.

The voltage V_2' is impressed on an *L* section and the output ripple is

$$V_{r,rms} = I_{rms}X_{c_2}$$
$$= \frac{V_2'}{X_L}X_{c_2}$$

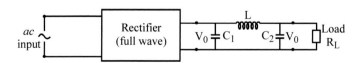

Fig. 2.26 CLC or π-type filter

Hence, the ripple factor is

$$\Gamma = \frac{V_{r,rms}}{V_{dc}} = \frac{V'_2 \frac{X_{c_2}}{X_L}}{V_{dc}}$$

$$= \sqrt{2}\,\frac{I_{dc}}{V_{dc}}\frac{X_{C_1}X_{C_2}}{X_L} \quad \text{But} \quad R_L = \frac{V_{dc}}{I_{dc}}$$

$$\Gamma = \sqrt{2}\,\frac{X_{C_1}X_{C_2}}{R_L X_L} \tag{2.26}$$

where all reactances are calculated for second harmonic frequency. If $f = 50$ Hz, capacitance in μF, L in H and R_L in ohms, then

$$\Gamma = \frac{5700}{LC_1C_2R_L} \tag{2.27}$$

i.e. inversely proportional to the inductor, capacitance as well as the load in π-filters.

Example 2.10 A π filter having $C = C_1 = C_2 = 10\ \mu$F and $L = 10\ H$ operates after a full-wave rectifier. The input voltage is $V = 300\ Sin(2\pi \times 50t)$ and the load resistance is $R_L = 10\ k\Omega$. Find the ripple factor (Γ) assuming the output voltage has only second harmonic. What is the ripple factor (Γ) if inductor is replaced by a resistance (R) equal to $10\ k\Omega$.

Solution

$$\Gamma = \sqrt{2}\,\frac{X_{c1} \times X_{c2}}{R_L X_L}$$

$$C = C_1 = C_2 = 10\ \mu F, \quad f = 50\ Hz, R_L = 10\ k\Omega$$

$$L = 10\ H$$

$$\Gamma = \frac{5700}{LC_1C_2R_L} = \frac{5700}{10 \times 10 \times 10 \times 10 \times 10^3}$$

$$= 5700 \times 10^{-7} = 0.0057\%$$

When inductor is replaced by the resistance $(R = 10\ k\Omega)$, then

$$\Gamma = \sqrt{2}\,\frac{1}{\omega^2 C_1 C_2 R_L R}$$

$$= \frac{\sqrt{2}}{(2\pi \times 50)^2 10 \times 10^{-6} \times 10 \times 10^{-6} \times 10 \times 10^3 \times 10 \times 10^3}$$

$$= 0.00036 = 0.036\%$$

Thus $10\ k\Omega$ service resistance gives lower ripples, but it will be at the cost of lower V_0.

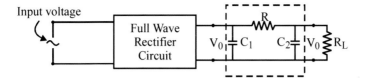

Fig. 2.27 RC filter

2.4.6 R–C Filler

On considering the CLC filter with the inductor L replaced by a resistance R, this type of filter is called R-C filter as shown in Fig. 2.27. The expression for the ripple factor can be obtained by replacing X_L in Eq. (2.26) by R giving

$$\Gamma = \sqrt{2}\frac{X_{c1}X_{c2}}{R_L R} \tag{2.28}$$

Therefore, if resistance R is chosen equal to the reactance of the inductor which it replaces, the ripple remains unchanged.

The resistance R will increase the voltage drop and hence, the regulation will be poor. This type of filters is often used for economic reasons, as well as the space and weight requirement of the iron-cored choke for the LC filter. Such R-C filters are often used only for low current power supplies as R is in series and reduces I_0 as well as V_0.

2.5   Diode Clippers

The circuit with which the waveform is shaped by removing (or clipping) a portion of the input signal without distorting the remaining part of the alternate waveform is called a clipper. Clipper is also referred to as voltage (or current) limiters, amplitude selectors, or slicers. These circuits find extensive use in digital electronic systems (like calculators, computers, radio and television receivers, etc.) and radar. Most of the clippers employ diodes in series or shunt and are known as diode clippers. Some of the importance of diode clippers are discussed below.

2.5.1 Positive Clippers

A circuit that resources the positive half cycles of the signal (input voltage) is called as a positive clipper. Such clipper as shown in Fig. 2.28 has all the positive half cycles clipped off (removed).

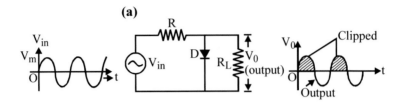

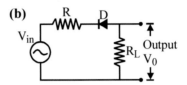

Fig. 2.28 a Positive clipper (shunt diode type) **b** Positive clipper (series diode type)

Working: During positive half cycle of input voltage in shunt diode-type positive differ (Fig. 2.28a), the diode D is forward biased and conducts heavily. Ideally, its acts as a short closed switch and hence the voltage across the diode or the load is zero and hence positive half cycle clipped off. In other words, the positive half cycle does not appear at the output.

During negative half cycle of input voltage, the diode D is reverse biased and behaves as an open switch. Then the current flows through R_L and series resistance (R). In this condition, the circuit behaves as a voltage divider, while the output voltage is taken across R_L.

$$\text{Then output voltage} = \frac{R_L}{R_L + R} \times V_m$$

$$\therefore \text{For negative half cycle, } V_m = -V_m(-\text{ve})$$

$$\text{So the output voltage} = -\text{negtagtive}$$

Above analysis is for positive clipper when the diode connected is shunt. Hence, it is also called as shunt-type positive clipper. Figure 2.28b shows also a positive clipper and here diode D is connected in series but in a reverse direction, operation of which is easily understandable.

2.5.2 Negative Clipper

In the negative clipper circuit, the diode D is connected in a direction opposite to that of positive clipper. In the series-type negative clipper as shown in Fig. 2.29a, during the positive half cycle of the input signal, the diode D conducts and acts as short circuit and hence, the positive half cycle of the input signal will appear at the output as shown in Fig. 2.29a.

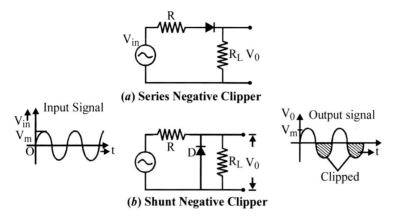

Fig. 2.29 a Series negative clipper, b shunt negative clipper

During the negative half cycle of the input signal shown in Fig. 2.29a, the diode D does not conduct and act as an open circuit. As a result, the negative half cycle will not appear at the output, i.e. the negative half cycle is clipped off i.e. removed (as shown in Fig. 2.29a, b)

Similarly, for shunt type of negative clipper Fig. 2.29b, the diode D at shunt does not conduct for positive half cycles and diode D behaves as open circuit. So, the output voltage will follow for given input signal for positive half cycle. During negative half cycle of the input signal, the diode D is forward biased and behaves as short circuit and hence no signal will appear across the diode D. So, the output voltage is clipped off during negative half cycle of the input signal. It is shown in Fig. 2.29b for shunt-type negative clipper.

It is evident from the above discussion that the negative clippers of series and shunt types behaves as half-wave rectifier. Thus, the negative clipper has clipped the negative half cycle completely and allow to pass the positive half cycle of the input signal.

2.5.3   Biased Positive Top Clipper

Figure 2.30 shows the circuits of shunt- and series-type positive clippers along with the input and output voltage waveforms. In the biased series positive clipper as shown in Fig. 2.30a, the diode D does not conduct as long as the input voltage is greater than $+V_R$ and hence, the output limited at $+V_R$. When the input voltage becomes less than $+V_R$, the diode becomes forward biased and conducts, acting as short circuit. Hence, all the input signal having less than $+V_R$ as well as negative half cycle of the input wave will appear at the output, as shown in Fig. 2.30c.

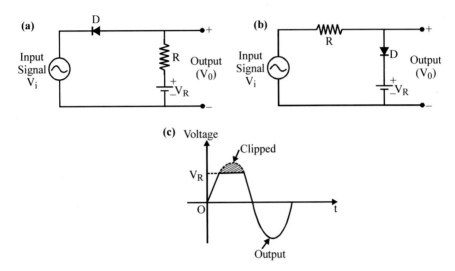

Fig. 2.30 Biased clipper **a** series diode type, **b** shunt diode type and **c** input and output signal waveforms

In the biased shunt positive clipper as shown in Fig. 2.30b, the diode D conducts as long as the input voltage is greater then $+V_R$ and the output remains at $+V_R$. When the input voltage is less than $+V_R$, the diode D does not conduct and acts as open switch. Hence, all the input signal having less than $+V_R$ as well as negative half cycle of the input wave will appear at the output, as shown in Fig. 2.30.

The clipping level can be shifted up or down by varying the bias voltage V_R.

2.5.4 Biased Positive Upper Clipper with −ve Polarity of the Battery (V_R)

Figure 2.31 shows the biased series- and shunt-type clippers with reverse polarity of V_R along with the input and output voltage waveforms. Here, the entire signal above $-V_R$ is clipped off.

2.5.5 Biased Negative Bottom Clipper

In the biased series negative clipper, as shown in Fig. 2.32a, when the input voltage$(V_i) < -V_R$, then the diode D does not conduct and clipping taken place. In the biased shunt type of clipper shown in Fig. 2.32b, when the input voltage $V_i \leq -V_R$ then the diode D conducts and clipping taken place. The clipping level can be shifted up and down, on varying the bias voltage $(-V_R)$.

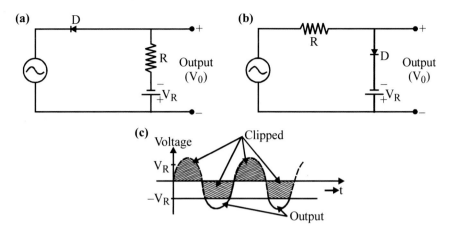

Fig. 2.31 Biased positive clipper with reverse polarity $(-V_R)$ **a** series, **b** shunt and **c** input and output signal waveforms

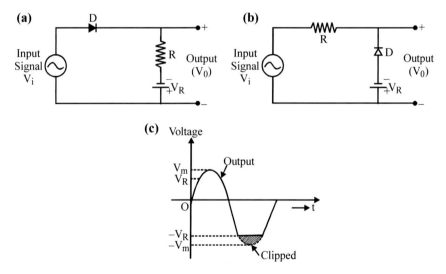

Fig. 2.32 Biased negative clipper **a** series type, **b** shunt type and **c** input and output signal waveforms

2.5.6 Biased Negative Lower Clipper with +ve Polarity of the Battery (V_R)

Figure 2.33 shows the biased series and shunt negative clippers with reverse polarity $(-V_R)$ along with the input and output waveforms, as shown in Fig. 2.33c. Here, the entire signal below $+V_R$ is clipped off.

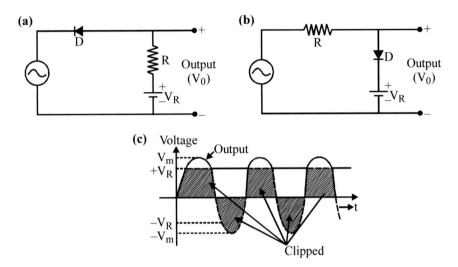

Fig. 2.33 Biased negative clipper with reverse polarity in V_R $(-V_R)$ **a** series, **b** shunt and **c** input and output signal waveforms

2.5.7   Combination Clipper (Two Biased Diodes Clipper)

Such type of clipper circuit is a combination of a biased positive clipper shunt type and a biased negative clipper shunt-type circuit. Figure 2.34 shown the combination clipper along with the input and output voltage waveforms. When the input signal voltage $V_i \geq + V_{R1}$ the diode D_1 conducts and acts as a closed switch, while diode D_2 is reverse biased and D2 acts as an open switch. Hence, the output voltage can exceed the voltage level of $+V_R$, during the positive half cycle.

Similarly, when the input signal voltage, $V_i \leq - V_{R2}$ the diode D_2 conducts and acts as a closed switch, while diode D_1 is reverse biased and D_1 acts as an open

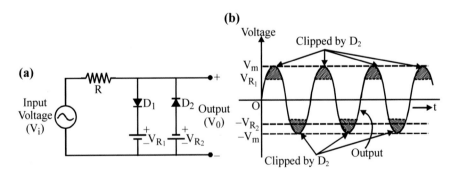

Fig. 2.34   a Combination clipping circuit and **b** input and output waveforms

switch. Hence, the output voltage V_0 cannot go below the voltage level of $-V_{R2}$ during the negative half cycle.

It is evident that the clipping level may be changed by varying the values of V_{R1} and V_{R2}. If $V_{R1} = V_{R2}$, the circuit will clip both the positive and negative cycles at the same voltage levels and hence, such a combination clipper is called symmetrical clipper.

2.5.7.1 Summary Table of Clippers

S. no.	Clipper circuit name	Circuit $\left\{ \begin{array}{l} V_i = Input \\ V_0 = Output \end{array} \right\}$	Input wave: solid and dotted line together Output wave: solid line only
1(a) 1(b)	Positive clipper (Series) Positive clipper (Shunt)		
2(a) 2(b)	Negative clipper (Series) Negative clipper (Shunt)		
3(a) 3(b)	Biased positive clipper (with series diode) Biased positive clipper (with shunt diode)		

(continued)

(continued)

S. no.	Clipper circuit name	Circuit $\left\{ \begin{array}{l} V_i = Input \\ V_0 = Output \end{array} \right\}$	Input wave: solid and dotted line together Output wave: solid line only
4(a) 4(b)	Biased positive clipper with reverse polarity (with series diode) Biased positive clipper with reverse polarity (with shunt diode)		
5(a) 5(b)	Biased negative clipper (with series diode) Biased negative clipper (with shunt diode)		
6(a) 6(b)	Biased negative clipper with reverse polarity of V_R (with series diode) Biased negative clipper with reverse polarity of V_R (with shunt diode)		
7.	Combination clipper (Both diodes in shunt type)		

Questions

Fill in the blanks

1. The breakdown mechanism in a lightly doped *pn* junction under reverse-biased condition is called _____ and in heavily doped is _____.
2. _____ diode is used in reverse-bias mode only.
3. A Zener diode has a sharp breakdown at low _____.
4. Aluminium is a _____ element where as phosphorus is a _____ element.
5. PIV of full-wave centre-tap rectifier has _____ V, whereas PIV of bridge rectifier has _____ V.

Short Questions

1. Give the V-I characteristics of *pn* junction diode and define PIV and cut-in voltage.
2. Give the V-I characteristics of Zener diode and Avalanche diode.
3. Define the terms.

   (i) Breakdown voltage, (ii) knee voltage, (iii) destructive thermal breakdown

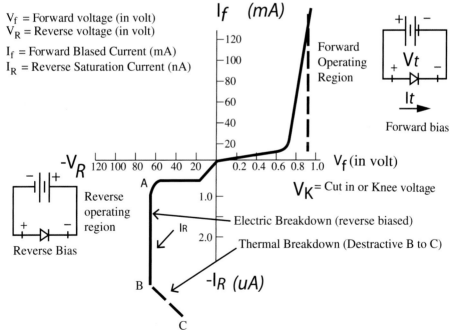

4. Define positive and negative clippers.

5. What is the effect of temperature on Zener diode.
6. Write short notes on the following half-wave rectifier, Bridge rectifier and full-wave centre-tap rectifier.
7. What is Zener diode? Why is Zener diode used as voltage regulator and not the Avalanche diode?
8. Define peak inverse voltage and maximum power rating.
9. Define Avalanche breakdown.
10. List down all the properties of various filters L, C, LC, CLC.
11. 'Depletion region is a space charge region'. Explain how? Why its resistance is quite high?

Long Questions

1. Explain the operation of *pn* junction diode. Draw the V-I characteristics of *pn* junction diode.
2. Explain potential barrier in unbiased and reverse biased of *pn* junction diode. What is transition capacitor?
3. Explain the Zener diode and draw its V-I characteristics.
4. Explain the working of bipolar junction transistors in active, cut-off and saturation regions.
5. Explain the action of *pn* junction diode under forward bias and reverse bias. Write down the application of *pn* junction diode.
6. Explain the mechanism of Avalanche breakdown and Zener breakdown.
7. Differentiate between Avalanche breakdown and Zener breakdown.
8. With the help of neat sketches, explain the various types of filters L, C, LC and CLC.
9. Differentiate between full-wave centre-tapped rectifier and bridge rectifier.
10. Reverse saturation current in an Si diode is 5 mA. Find the voltage across it at 500 mA at 27 °C.
11. Draw I-V characteristics of Si and Ge on same graph for comparison.
12. Draw I-V characteristics of *pn* junction diode for temperature T_1 and T_2 where $T_2 > T_1$.
13. 'Depletion width depends on reverse bias and doping densities of p and n sides of *pn* junction diode'. Explain with expressions.
14. Draw band structure diagram of *pn* junction diode with and without reverse bias. What is the potential barrier in each case?
15. Explain about the diffusion capacitor and transition capacitor?
16. Explain static and dynamic resistance of a diode in forward- and reverse-biased diode. Write the range of the values for Si diode.
17. For (a) half-wave rectifier and (b) full-wave rectifier, the transformer with the ratio of primary and secondary turns is 4:1. The primary voltage is 220 V and load resistance is 50 Ω. On assuming the ideal diode, find the ripple factor, RMS voltage, *dc* power, *dc* voltage, efficiency, and frequency of ripples at output for each rectifier.

18. 'For a *pn* junction diode, the reverse-bias current is due to minority carrier'. Explain.
19. Explain the functioning of diode as an ON/OFF switch.
20. Explain the formation of capacitor in forward- and reverse-biased *pn* junction diode. In this capacitor what is the dielectric?
21. Write the expression for efficiency, regulation and TUF in rectifiers.

Chapter 3
Junction Transistors and Field-Effect Transistors

Contents

© Springer Nature Singapore Pte Ltd. 2020
S. S. Srikant and P. K. Chaturvedi, *Basic Electronics Engineering*,
https://doi.org/10.1007/978-981-13-7414-2_3

3.1 Introduction of Bipolar Junction Transistors (BJT) and Its Construction

Transistor was first invented in 1948 by John Bardeen and WH Brattain of Bell Laboratories, which was commercially used in the telephone circuits in 1951.

A bipolar junction transistor (BJT) has three differently doped semiconductor regions. Two of these regions are doped with either acceptor or donor atoms, and the third region is doped with another type of atoms. Actually, a BJT consists of two PN junctions which are virtually back to back. The word 'bipolar' is used to state the role of both charge carriers (free electrons and holes).

So, one can define the transistor as 'A semiconductor device consisting of two pn junctions and looks like by sandwiching either *p*-type or *n*-type semiconductor between a pair of opposite types'.

Accordingly, there are two types of transistors, namely,

(i) *npn* transistor and (ii) *pnp* transistor.

(i) **npn transistor**: A transistor in which two blocks of *n*-type semiconductor are separated by a thin base layer of *p*-type semiconductor is called *npn* transistor (Fig. 3.1a).

(ii) **pnp transistor**: A transistor in which two blocks of *p*-type semiconductor are separated by a thin base layer of *n*-type semiconductor is called *pnp* transistor (Fig. 3.1b).

Every transistor has three terminals: emitter (E), base (B) and collector (C). The symbolic representation of the two types of BJT is shown in Fig. 3.2. The arrow on the emitter specifies the direction of current flow when the EB junction is forward biased, just like that in pn junction diode.

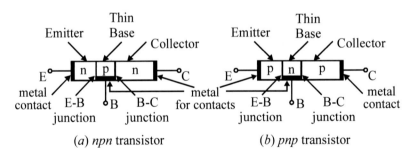

Fig. 3.1 **a** *npn* transistor, **b** *pnp* transistor

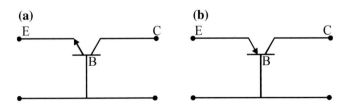

Fig. 3.2 Circuit symbol **a** *npn* transistor, **b** *pnp* transistor

Emitter is heavily doped (10^{17}/CC) so that it can inject a large number of charge carriers into the base. Base is moderately doped (10^{15-16}/CC) and is very thin; it therefore passes most of the injected charge carriers from the emitter into the collector, which is most lightly doped (10^{13-14}/CC).

3.2 Transistor Biasing

When a transistor is unbiased, the free electrons diffuse across the junction forming two depletion layers (see Fig. 3.3). Due to this, charge accumulation in these depletion layers takes place and therefore barrier potential V_o is developed, which is nearly 0.7 V at 27 °C (room temperature) for a silicon transistor (0.3 V for a germanium transistor). The width of the two depletion layers across each junction will be of different widths as the three regions are doped at different levels. The more heavily doped a region is, the greater is the concentration of charge near the junction. This means that the depletion layer penetrates less into the emitter region (heavily doped) but deeply into the base region (moderately doped). Therefore, the total depletion width formed across the EB junction is smaller (as shown in Fig. 3.3).

However, across the collector junction, the depletion layer penetrates into the base (moderately doped) to some extent and quite a deep into the collector most lightly doped). Therefore, the total width of the depletion layer formed across the collector junction is larger comparatively as shown in Fig. 3.3. This summarizes that the depletion layer at the emitter junction is small and the depletion layer width at the collection junction is larger even at zero bias.

When a transistor is biased and active, normally the emitter–base (EB) junction is forward biased and collector–base junction is reverse biased. Then

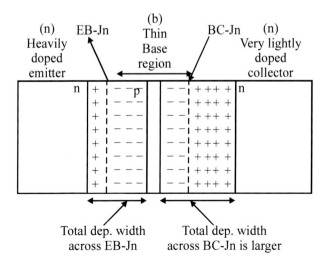

Fig. 3.3 Unbiased *npn* transistor depletion layers formed across an EB junction and BC junction, due to diffusion of charges across junctions

two thing happens: (a) Due to the forward bias on the emitter–base junction, the EB depletion width and barrier potential +0.7 V first get cancelled and then only emitter supplies or injects large amount of majority carriers into the base, as emitter is heavily doped. (b) This brings the majority carrier coming from emitter more closer to the electric field of reverse-biased CB junction. This field attracts most of the majority carriers coming from emitter to base into the collector side, before recombination of these carriers in the thin base takes place. For facilitating these two actions, the base is made thin. **This is what we call as the transistor action due to these two reasons**. We will see later that a small change in base current (μA) in BJT changes collector current (mA) significantly and therefore we call it a current-controlled amplifier/device.

In the common base configuration, the base–emitter junction is forward biased, thus normally providing low resistance for the emitter circuit and the base–collector junction is reverse biased, offering high resistance path to the collector circuit (Fig. 3.4a). The base is made very thin, so that it can pass on most of the majority carriers supplied by the emitter to the collector.

Since there are two junctions in a transistor, namely, emitter–base junction and collector–base junction, each of these two junctions may be forward biased or reverse biased. Therefore, there are four possible ways of biasing these two junctions (Fig. 3.4). These possible ways are also called as modes of operation of a transistor. These different modes are listed in Tables 3.1 and 3.2.

> ***Important***: *A bipolar transistor is frequently used as a constant current source.*

Case (a): Active region (mode): In this mode, the emitter–base junction is forward biased and collector–base junction is reverse biased.

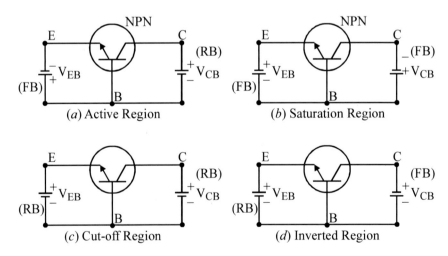

Fig. 3.4 The four modes of operation of a *npn* transistor

Table 3.1 Modes of operation of BJT

Cases	Emitter–base Junction	Collector–base Junction	Region mode of Operation
a	Forward biased (FB)	Reverse biased (RB)	Active mode
b	Forward biased (FB)	Reverse biased (RB)	Saturation mode
c	Reverse biased (RB)	Reverse biased (RB)	Cut-off mode
d	Reverse biased (RB)	Forward biased (FB)	Inverted mode

Table 3.2 Modes of operation of a transistor depends on the direction of biasing of BE and BC junctions

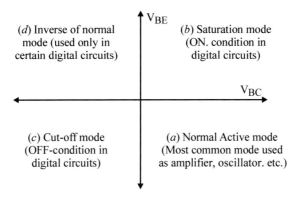

(d) Inverse of normal mode (used only in certain digital circuits)

(b) Saturation mode (ON. condition in digital circuits)

(c) Cut-off mode (OFF-condition in digital circuits)

(a) Normal Active mode (Most common mode used as amplifier, oscillator. etc.)

The battery V_{EB} is connected between emitter and base, in order to make emitter–base junction forward biased. Similarly, the battery V_{CB} is connected between collector and base in order to make collector–base junction reverse biased. In this mode, the transistor is used for amplification. In this mode, the collector current depends upon the input current I_E. It is shown in Fig. 3.4a.

Case (b): Saturation region (mode): In this mode (region), both emitter–base junction and collector–base junction are forward biased. In this mode, the collector current becomes independent of the input current and therefore, the transistor acts like a closed switch. It is shown in Fig. 3.4b.

Case (c): Cut-off region (mode): In this mode (region), both the junctions are reverse biased. In this region, the transistor has zero current because the emitter does not emit charge carriers into the base and no charge carriers are collected by the collector except a few thermally generated minority carriers. In this mode, the transistor is open when it is used as an open switch. It is shown in Fig. 3.4c.

Case (d): Inverted region (mode): In this region (mode), the emitter–base junction is reverse biased and the collector–base junction is forward biased. Since the doping level of emitter and collector is not the same, therefore, the collector cannot inject much majority charge carriers into the base. *In this mode, the transistor action is poor,* and hence this region is of little importance. It is shown in Fig. 3.4d.

3.3 Working of *npn* and *pnp* Transistor

3.3.1 npn *Transistor*

The transistor with *npn* configuration is shown in Fig. 3.5. The emitter–base junction is forward biased, while collector–base is reverse biased. The forward-biased voltage V_{EB} is quite small but larger than barrier potential of 0.7 V

(a) with $\quad I_E = I_{En} + I_{Ep}; I_c = I_{cn} + I_{cp}; I_B = I_{Bn} + I_{Bp};$

$\qquad\qquad\qquad I_E = I_C + I_B$

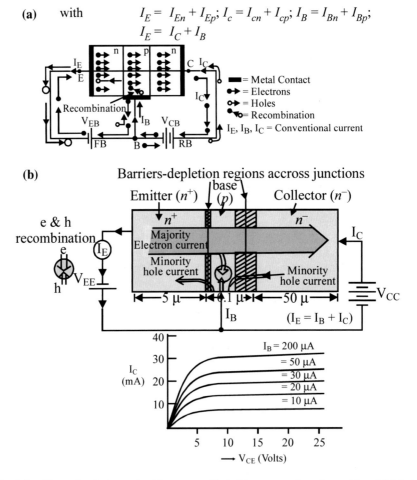

Fig. 3.5 a Flow of majority carrier (electrons and holes) in *npn* transistor J_{EB} and J_{BC}. **b** Majority carrier current in the three regions of *npn* transistor, 95% of electrons from emitter reach collector and 5% constitute base current. Minority carriers in these three regions are much smaller than majority. In *pnp* transistor, same figure with changed bias direction is true with 'e' replaced by 'h'

and reverse-biased voltage V_{CB} is considerably high, but smaller than breakdown voltage.

As the emitter–base junction is forward biased, therefore (a) a large number of free electrons (majority carriers) in the emitter (*n*-type region) are pushed towards the base. This constitutes the emitter current I_E. When these electrons enter the *p*-type material (base), they tend to combine with holes. Since the base is lightly doped and very thin, only a few electrons (less than 5%) combine with holes to constitute base current (I_B) (b). The remaining free electrons (more than 95%) diffuse across the thin base region and get attracted to the collector having positive space charge layer with high *E*-field and get collected by the collector. **These two reasons lead majority carriers of I_E to constitute collector current (I_C). *This process is the transistor action* (see Fig. 3.5b)**. Thus, it is seen that almost the entire emitter current (I_E) flows into the collector circuit. However, to be more precise, the emitter current (I_E) is the sum of current (I_C) and base current (I_B), with each constituting of holes and electron current, *i.e.*

$$\text{with} \quad I_E = I_{En} + I_{Ep}; I_c = I_{cn} + I_{cp}; I_B = I_{Bn} + I_{Bp};$$
$$I_E = I_C + I_B$$

3.3.2 pnp *Transistor*

The transistor with *pnp* configuration is shown in Fig. 3.6. Here, the mechanism is nearly the same as in *npn* except that the majority carrier here is hole as dominant carrier. The emitter–base junction is forward-biased voltage V_{EB}, which is quite smaller but greater than barrier potential V_b, whereas the reverse-biased voltage V_{CB} is considerably high, but smaller than the breakdown voltage.

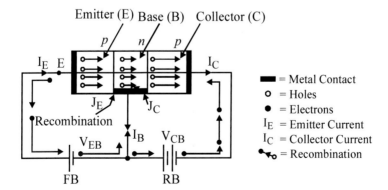

Fig. 3.6 Flow of majority carriers in *pnp* transistor

As the emitter–base junction is forward biased, therefore (a) a large number of holes (majority carriers) in the emitter (*p*-type semiconductor) are pushed towards the base. This constitutes the emitter current (I_E). When these holes enter the *n*-type material (base), they tend to combine with free electrons. Since the base is lightly doped and is very thin, only a few holes (less than 5%) combine with free electrons to constitute base current (I_B) (b). The remaining holes (more than 95%) diffuse across the thin base region and reach the collector space charge layer. These holes then come under the influence of the negatively biased *p* region and are attracted or collected by the collector. This constitutes collector current (I_C). Thus, it is observed that almost the entire emitter current flows into the collector circuit. However, to be more precise, the emitter current is the sum of collector current and base current, *i.e.*

$$I_E = I_C + I_B$$

It may be noted that inside the transistor, the current is constituted by the change of position of the holes and electrons, whereas, outside the transistor, *i.e.* in the leads of the circuit, the current is constituted by the flow of electrons (also called as conventional current).

Important: **Most of the transistors used are *npn* type and not *pnp* type** because in *npn* transistors, the current conduction is mainly by electrons, whereas in *pnp* transistor, the current conduction is mainly by holes. The electrons are more mobile than holes, leading to higher conduction in *npn* transistor and higher gain, etc.

3.3.3 Minority Carrier and Potential Inside **pnp** Transistor

As we know that when the emitter–base (EB) junction is forward biased and the collector–base (CB) junction is reverse biased, then the minority carrier concentration will be different from the thermally equilibrium minority carrier concentration at zero bias. Due to injected carriers across the junction J_E, the hole concentration p_n in the base and the electron concentration n_p in the emitter near junction J_E are increased significantly as shown in Fig. 3.7 for *pnp* transistor. The value of p_n and n_p decreases with increasing distance from junction J_E. The hole density p_n becomes zero at the collector–base junction (J_C). Similarly, the concentration of n_p at collector–base junction (J_C) becomes zero by recombination due to reverse bias of collector–base junction.

The potential variation of a *pnp* transistor at biased condition is shown in Fig. 3.8 on taking the reference from Fig. 3.6. The dotted curve shows the potential before applying the bias potential. It is evident from Fig. 3.8 that the forward biasing of the emitter junction decreases the emitter–base potential barrier V_o by V_{EB}. The reverse biasing of the collector junction increases the collector–base

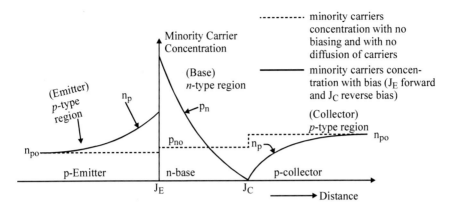

Fig. 3.7 Minority carrier concentrations in a *pnp* transistor under unbiased and biased condition

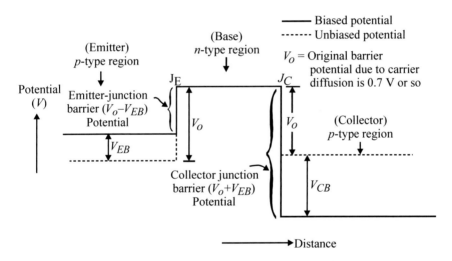

Fig. 3.8 Potential of a *pnp* transistor at biased condition

potential by V_{CB}. Therefore, emitter junction barrier voltage is $V_o - V_{EB}$ and the collector junction barrier voltage is $V_o + V_{CB}$.

In Fig. 3.8, the space charge regions on the junction J_E and J_C are neglected. When the width of the depletion layer is present at J_E and J_C junction, Fig. 3.8 will be modified. The width of depletion layer of a PN junction is increased by increasing the reverse-bias voltage, but the depletion layer width decreases with increasing the forward-bias voltage. As the emitter–base junction (J_E) is forward biased, the depletion layer or barrier width of J_E is neglected. Since the collector–base junction (J_C) is reverse biased, the depletion layer or space charge width increases and exists at junction J_C.

3.4 Early Effect or Base-Width Modulation

The transition region at junction J_C has a region of uncovered charge on the both sides of the junction. When the reverse bias voltage across the junction increases, the transition region can penetrate deeper into the base and collector region as shown in Fig. 3.9.

To retain the neutrality of charge, the net charge must be zero and the number of uncovered charge on each side must be equal. Since the doping concentration in the base region is comparatively larger than in the collector, the penetration of depletion layer or space charge region into the base region is smaller as compared to the space region in collector region. But the depletion layer width in the base region cannot be neglected (shown in Fig. 3.9), as for transistor action base is kept very thin.

When the metallurgical base width is W_B and the space charge width (W) is the width of transition region due to depletion, the effective electrical base width becomes $W'_B = W_B - W$. **The modulation of the effective base width due to reverse bias voltage across the collector–base junction (J_C) is called early effect or base width modulation.** The reduction in effective base width with increasing reverse collector–base voltage has the following consequences.

(i) There is a very small possibility up to 5% for recombination of electrons and holes in the base region.
(ii) The concentration gradient of minority carrier (p_n) increases within the base. The hole concentration (p_n) is zero at a distance (d), which is in between W_B and W'_B, but the potential at J_C becomes below V_o. Figure 3.10 shows the

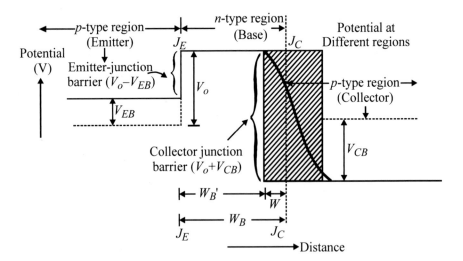

Fig. 3.9 Potential of a *pnp* transistor with space charge width (W) inside base

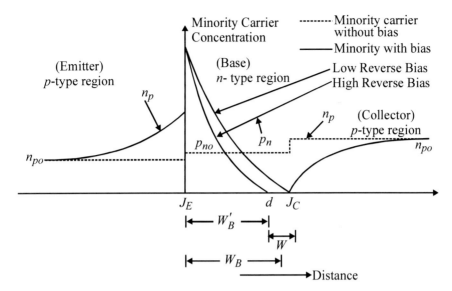

Fig. 3.10 Minority carrier density in the base region without bias and at low reverse bias and high reverse bias at collector

   minority carrier density within base region at both low and high reverse bias conditions.

(iii)   The hole current injected across the emitter is proportional to the gradient of p_n at J_E and emitter current (I_E) increases with increasing collector reverse base voltage. Also, I_c keeps increasing with V_{CB} (Fig. 3.14) as a result $\propto$ is also not constant as $\alpha = (\partial I_C / \partial I_B)$.

(iv)   At very high voltage reverse voltage of collector the effective electrical base width (W'_B) becomes zero, causing voltage breakdown in the transistor. This phenomenon is called **punch through**.

   The above process is Early Effect phenomena for *pnp* transistor; similarly, one can define base width modulation or Early Effect for *npn* transistor also.

3.5   BJT Configuration and Characteristics

When a bipolar junction transistor (BJT) is to be connected in a two-port network circuit as shown in Fig. 3.11a, then one terminal (port) is used as an input terminal, the other terminal (port) is used as an output terminal and the third terminal is made common between input and output terminals.

   Depending upon the input, output and common terminals, a transistor can be connected in three configurations.

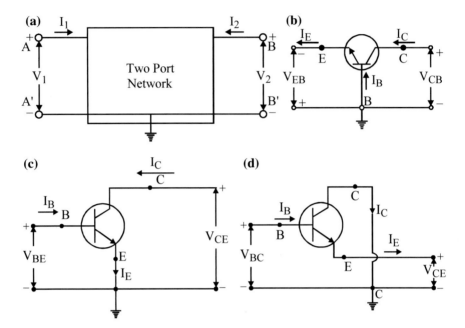

Fig. 3.11 a A general two-port network. **b** A BJT transistor (CB configuration) as a two-port network (*npn* transistor). **c** CE configuration (using *npn* transistor). **d** CC configuration (using *npn* transistor)

(a) Common base (*CB*) configuration as shown in Fig. 3.11b.
(b) Common emitter (*CE*) configuration as shown in Fig. 3.11c.
(c) Common collector (*CC*) configuration as shown in Fig. 3.11d.

It is important to note here that the transistor may be connected in any one of the above-said three configurations, but emitter–base junction is always forward biased and collector–base junction is always reverse biased to operate the transistor in active region.

3.5.1   Common Base (CB) Configuration

The common base (CB) configuration for *npn* transistor and *pnp* transistor with biasing is shown in Fig. 3.12a and 3.12b, respectively.

Finding input–output characteristic

In common–base configuration, two types of characteristics are used:

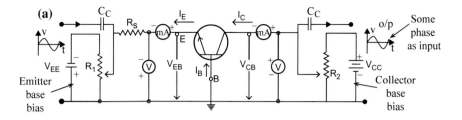

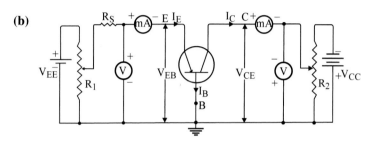

Fig. 3.12 **a** CB configuration for *npn* transistor biased to work in active region as signal amplifier with output in the same phase as input with no amplification. **b** CB configuration for *pnp* transistor biased to work in active region

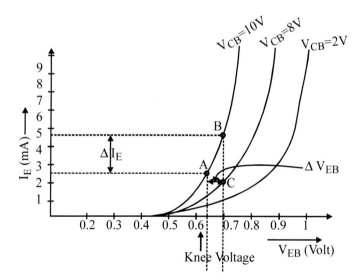

Fig. 3.13 Input characteristics in CB configuration

(a) **Input characteristics**: Input characteristic curves relate the input or emitter current I_E with collector-to-base voltage (V_{CB}) constant. The input characteristics curves shown in Fig. 3.13 are obtained for *npn* transistor as shown in

Fig. 3.12a. The series resistance R_S is connected to limit the emitter current (I_E). The value of voltage V_{EB} may be varied with the help of a potentiometer (R_1). The collector voltage (V_{CB}) is varied adjusting the potentiometer R_2. The *dc* milliammeter and *dc* voltmeter are connected in the emitter and collector circuits (as shown in Fig. 3.12a) to measure the voltages and currents.

The input characteristics in CB configuration are plotted between emitter current (I_E) and emitter–base forward bias voltage (V_{EB}), for different values of collector–base voltage (V_{CB}) (as shown in Fig. 3.12a). A number of characteristic curves can be plotted for different settings of V_{CB}.

Figure 3.13 shows the input characteristics of a typical *npn* transistor in common base configuration. The following points may be noted from these characteristics:

(i) For a particular value of V_{CB}, the curve is just like a diode characteristics in the forward region, as here the *pn* emitter junction is forward biased.
(ii) When V_{CB} increases from 8 V to 10 V (Fig. 3.13) from A to B, the emitter current I_E increases rapidly 2.5–4.5 mA with a small increase in emitter–base voltage V_{EB} 0.64–0.70 V.

The ratio of change in emitter–base voltage (ΔV_{EB}) to the resulting change in emitter current (ΔI_E) at constant collector–base voltage (V_{CB}) is known as input resistance, *i.e.*

$$r_i = \Delta V_{EB}/\Delta I_E = (0.70 - 0.64)/(4.5 - 2.5) \times 10^{-3} = 30\ \Omega \qquad (3.1)$$

This shows that the input resistance is very small. The value of input resistance r_i further decreases with the increase in collector–base voltage V_{CB} since the curve

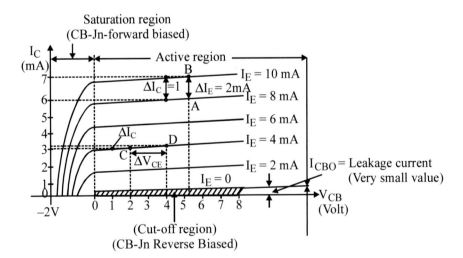

Fig. 3.14 Output characteristics in a CB configuration for typical *npn* transistor

tends to become more vertical (see Fig. 3.13). The typical value of input resistance varies from a few ohms to 100 Ω.

(b) **Output characteristics**: Output characteristics curve relate the output or collector current I_C and output or collector-to-base reverse bias voltage V_{CB} keeping the input or emitter current I_E constant.

A number of characteristic curves can be plotted for different settings of I_E. Figure 3.14 shows the output characteristics of a typical *npn* transistor in *CB* configuration. The following points from these characteristics are worth noting:

(i)   The characteristics curve show the three regions, namely, active region, cut-off region and saturation region.

(ii) In active region, the collector current is approximately equal to the emitter current I_E.

(iii) If V_{CB} becomes negative, the collector–base junction is set in forward bias which causes collector current to decrease rapidly. This region is known as saturation region. In saturation region, the collector current does not depend much on emitter current.

(iv) In saturation region, since collector–base junction is also forward biased, a small increase in V_{CB} results in a large increase in collector current.

(v)   If emitter current I_E is made zero, then the collector current is not zero but has a very small value of leakage current. In this region, both the junctions are reverse biased.

(vi) Emitter current increases very slowly for increases of V_{CB} in active region; hence, the curves are nearly flat. This also means that the *dc* output resistance in this configuration is very large (≈ 1.5 KΩ).

(vii) With the help of output characteristics Fig. 3.14, the dynamic output resistance r_0 can be found by the expression for $I_E = 2$ mA (point *C–D*):

$$r_0 = \frac{\Delta V_{CB}}{\Delta I_C} \text{ at constant } I_E = \left[\frac{4 - 2}{(3 - 2.8 \times 10^{-3})}\right] \approx 10 \text{ K}\Omega$$

Here, ΔV_{CB} and ΔI_C are small changes in collector-to-emitter voltage and collector current for given emitter current $I_E = $ Constant $= 4$ mA from points *C* to *D*.

(viii) With the help of output characteristics curve *dc* and *ac* current gains α_{dc} and α_o in CB configuration can also be determined as follows:
dc current gain is (at point *A*)

$$\alpha_{dc} = \alpha = \frac{I_C}{I_E} \approx \frac{6}{8} \approx 0.75 \tag{3.3}$$

where I_C and I_E are the values of collector and emitter currents at any point on the curve.

The *ac* current gain (Point *A–B*):

$$\alpha_0 = \frac{\Delta I_C}{\Delta I_E} \left(\text{at constant } V_{CB} \text{ of 5.3 V}\right) \approx \frac{7-6}{10.8} = 0.5 \qquad (3.4)$$

3.5.2 Common–Emitter (CE) Configuration

The common–emitter (CE) configuration for *npn* transistor and *pnp* transistor with proper biasing are shown in Fig. 3.15a and 3.15b, respectively.

In a common–emitter (*CE*) configuration, a transistor has two types of characteristics, namely, input characteristics and output characteristics.

1. **Input characteristics**: In *CE* configuration, the curve plotted between base current I_B and base–emitter voltage V_{BE} at constant collector–emitter voltage V_{CE} is called input characteristics.

   To draw the input characteristics, we note down the reading of ammeter (I_B) connected in the base circuit for various values of V_{BE} at constant V_{CE}. Plot the curve on the graph taking I_B along *y*-axis and V_{BE} along *x*-axis as shown in

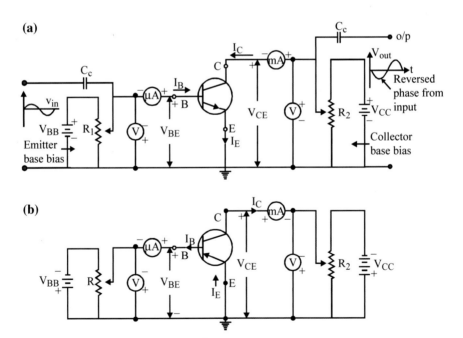

Fig. 3.15 CE configuration as signal amplifier, with phase reversal of *ac* from input for **a** *npn* transistor, **b** *pnp* transistor

Fig. 3.16 Input
characteristics of a *npn*
transistor for CE
configuration

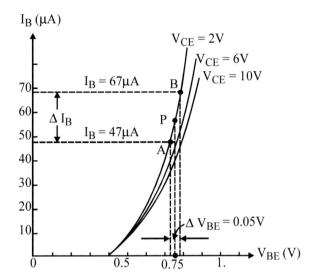

figure. Then draw a family of curves for different settings of V_{CE}. The following
points are worth noting from these characteristics:

(i)   These curves are similar to those obtained for *CB* configuration, *i.e.* like a
      forward diode characteristics. The only difference is that in this case I_B
      increases less rapidly with increase in V_{BE}. Hence, the input resistance of
      CE configuration is comparatively higher than that of *CB* configuration.

(ii) The change in V_{CE} does not result in a large deviation of the curves and
      hence, the effect of change in V_{CE} on the input characteristics is ignored for
      all practical purposes.

      Input resistance: The ratio of change in base–emitter voltage (ΔV_{BE}) to
      the resulting change in base current (ΔI_B) at constant collector–emitter
      voltage (V_{CE}) is known as input resistance, *i.e.*

$$\boxed{r_i = \frac{\Delta V_{BE}}{\Delta I_B}} \text{ at constant } V_{CE} \qquad (3.6)$$

In Fig. 3.16, if constant $V_{BE} = 0.75$ V around the point P, then for *A–B*:
$$r_i = \Delta V_{BE}/\Delta I_B$$
$$= 0.05/(67 - 47) \times 10^{-6} \ \Omega$$
$$= 0.005/2 \times 10^{-6} \ \Omega$$
$$= 2.5 \ \text{K}\Omega$$

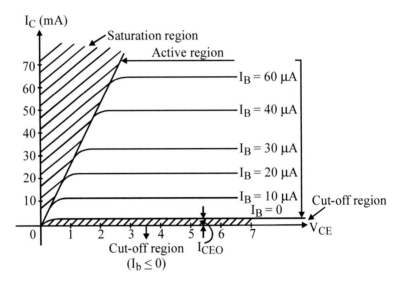

Fig. 3.17 Output characteristics in CE configuration

2. **Output characteristics**: In *CE* configuration, the curve plotted between collector current I_C and collector emitter voltage V_{CE} at constant base current I_B is called output characteristics. Figure 3.17 shows the typical output characteristics. The following points are worth noting from these characteristics.

   (i)   The output signal gets reversed (180° out of phase) (see Fig. 3.15a)
   (ii) Since the value of I_C increases with the increase in V_{CE} at constant I_B, the value of β also increases (as β = I_C/I_B).
   (iii) When V_{CE} falls below the value of V_{BE} (*i.e.* below a few tenths of a volt), I_C decreases rapidly. In fact, at this stage, the collector–base junction enters the forward-biased region and the transistor works in the saturation region. In the saturation region, I_C becomes independent and it does not depend upon the input current I_B.
   (iv) In the active region, $I_C = \beta I_B$, **hence, a small change in base current I_B in µA produces large change in output current (I_C) in mA and this is the secret of a *CE* amplifier, and hence BJT is called a current-controlled device.**
   (v)   If base current is made zero, then the collector current (I_E) is not zero. It has a value known as reverse leakage current I_{CEO}. This condition is called the **cut-off region of transistor**.
   (vi) In the active region, I_C increases slightly as V_{CE} increases. The slope of the curve is little bit more than the output characteristics of CB configuration. Hence, the *dc* output resistance of this configuration is slightly less as compared to *CB* configuration, but of the order of 0.5 KΩ.

(vii) The output characteristics may be used to find the dynamic output resistance r_o at any given point and has value in the order of 50 KΩ (very high).

$$r_o = \frac{\Delta V_{CE}}{\Delta I_C}\bigg|_{I_B=\text{constant}} \tag{3.7}$$

(viii) The output characteristics may also be used to calculate the *dc* current gain β and *ac* current gain $β_0$ as follows.

$$dc \text{ current gain } β = \frac{I_C}{I_B} \tag{3.8}$$

$$ac \text{ current gain } β_0 = \frac{\Delta I_C}{\Delta I_B}\bigg|_{V_{CE}=\text{constant}} \tag{3.9}$$

Here, also the voltage gain will be there (v_o/v_i) as amplifier (as $r_o \gg r_i$), but with phase reversal as in Fig. 3.15.

Example 3.1 The output characteristics of an *npn* transistor in CE configuration are shown in Fig. 3.18. Determine for this transistor.

(i) The dynamic output resistance,
(ii) The *dc* current gain and
(iii) The *ac* current gain at an operating point $V_{CE} = 10$ V when $I_B = 40$ μA.

Ans. For $V_{CE} = 10$ V, $I_B = 40$ μA.

The operating Q is marked in Fig. 3.18. At the operating point Q, the collector current $I_C = 4.55$ mA.

To determine the dynamic output resistance, let the small change in V_{CE} from A to B around the operating point Q be from 7.5 V to 12.5 V. The change in V_{CE} is

$$\Delta V_{CE} = 12.5 - 7.5 = 5 \text{ V}$$

The corresponding change in I_C at constant $I_B = 40$ μA will be

$$\Delta I_C = 4.7 - 4.3 = 0.4 \text{ mA}$$

$$\therefore \quad r_o = \frac{\Delta V_{CE}}{\Delta I_C} \text{ at constant } I_B = 40 \, \mu A$$

$$= \frac{5V}{0.4 \, mA} = \frac{5}{0.4 \times 10^{-3}}$$

$$= 12.5 \text{ KΩ}$$

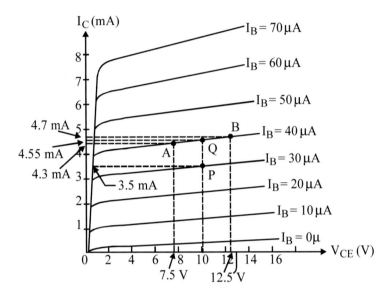

Fig. 3.18 Example 3.1: Output characteristic of CE configuration of a *npn* BJT

To find β_{dc},

$$\beta_{dc} = \frac{I_C}{I_B} = \frac{4.55}{40\mu A} = 113.75$$

In order to find *ac* current gain (β_o), *i.e.* $\beta_o = \frac{\Delta I_C}{\Delta I_B}$, we draw a vertical line PQ corresponding to $V_{CE} = 10$ V. From the given characteristic, it is clear that when base current I_B changes from 30 μA to 40 μA, collector current correspondingly changes from 3.5 mA (point P) to 4.55 mA (point Q).

$$\therefore \quad \beta_o = \frac{\Delta I_C}{\Delta I_B}, \quad \text{at constant } V_{CE} = 10 \text{ V}$$

$$= \frac{(4.55 - 3.50)\,\text{mA}}{(40 - 30)\mu A} = \frac{1.05 \times 10^{-3}}{10 \times 10^{-6}} = 105$$

3.5.3   *Common Collector (CC) Configuration*

In the common collector (CC) configuration, collector is grounded for both *npn* and *pnp* transistor cases as shown in Fig. 3.19, respectively. In this arrangement, the input is connected between base and collector, while output is taken across the emitter and collector. Thus, the collector of the transistor is common to both input and output circuits and hence the name common collector configuration.

Current amplification factor (γ)

The ratio of output current to input current is called current amplified factor. In a common collector connection, the output current is emitter current (I_E), whereas the input current is base current (I_B).

Thus the ratio of change in emitter current to the change in bias current is known as current amplification factor. It is generally represented by Greek letter γ (Gamma),

i.e.

$$\boxed{\gamma = \frac{\Delta I_E}{\Delta I_B}} \tag{3.10}$$

It is seen that current amplification factor (γ) of common collector (CC) configuration is almost equal to current amplification (β) of common emitter (CE) configuration. This means that the value of current gain (γ) in CC configuration is very high and is comparable to the current gain (β) in *CE* configuration. However, this circuit in CC configuration is rarely used for amplification purpose because the voltage gain is very low (less than 1). The common collector (CC)

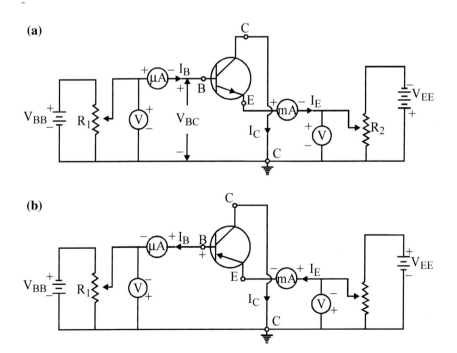

Fig. 3.19 Common collector configuration (CC) for **a** *npn* transistor, **b** *npn* transistor, *ac* input & *ac* input are in phase difference 180°

configuration is mainly used for impedance matching as Z_{in} is low but Z_o = very high over 600 KΩ, resulting in *ac* voltage gain <1, with no phase reversal of input signal.

3.5.4 Comparison Between CB, CE, CC Configurations

It has been observed that a transistor can be connected in any one of the three configurations, *i.e.* CB, CE and CC configuration. Their behaviours are different for different configurations. A configuration may be suitable for a particular application, but it may not be suitable for another.

Some important parameters which govern the suitability of a configuration are input dynamic resistance, output dynamic resistance, *dc* current gain, *ac* current gain, *ac* voltage gain and leakage current of the transistor.

For example, out of three configurations, the common collector configurations have very high input impedance but small value of voltage gain. So it is best suited for impedance matching the CE configuration.

Hence, to know the characteristics of three configurations at a glance, one should know their comparison and Table 3.3 shows the comparison.

Table 3.3 Comparison between the three configurations

S. no.	Characteristics	CB	CE	CC
1.	Input resistance (Ω)	Very low (100 Ω)	Low (100 KΩ)	Very high (750 KΩ)
2.	Output resistance (Ω)	Very high (500 KΩ)	High (10 KΩ)	Low (50 Ω)
3.	Current gain (A_I)	Less than unity (0.9)	High (100)	High (100)
4.	*ac* Voltage gain (A_V) phase	Small (150) (In phase)	High (500) (Phase reversed)	Less than unity (phase reversal)
5.	Leakage current	Very small (5μA for Ge and 1 μA for Si)	Very large (500 μA for Ge 20 μA for Si)	Very large (500 μA for Ge and 20 μA for Si)
6.	Applications	For high-frequency applications due to high A_v as at H.F. A_v is more important	For audio frequency amplifier/oscillator due to very high gain A_I	For impedance matching due to very high Z_{in} and very low Z_0.
7.	Signal phase reversal	No	Yes	Yes

3.5.5 Relation Among Current Amplification Factor of CB, CE and CC Configurations

In a transistor amplifier with *ac* input signal, the ratio of change in output current to the change in input current is called the current amplification factor.

In the *CB* configuration, the current amplification factor, $\alpha = \frac{\Delta I_C}{\Delta I_E}$

In the CE configuration, the current amplification factor, $\beta = \frac{\Delta I_C}{\Delta I_B}$

In the CC configuration, the current amplification factor, $\gamma = \frac{\Delta I_E}{\Delta I_B}$

Relationship between α and β: We know that $\Delta I_E = \Delta I_C + \Delta I_B$

By definition,

$$\Delta I_C = \alpha \Delta I_E$$

Therefore,

$$\Delta I_E = \alpha \Delta I_E + \Delta I_B$$

i.e.

$$\Delta I_B = (1 - \alpha)\Delta I_E$$

Dividing both sides by ΔI_C, we get

$$\frac{\Delta I_B}{\Delta I_C} = \frac{\Delta I_E}{\Delta I_C}(1 - \alpha)$$

Therefore

$$\frac{1}{\beta} = \frac{1}{\alpha}(1 - \alpha)$$

$$\boxed{\beta = \frac{\alpha}{1 - \alpha}} \tag{3.11}$$

Rearranging, we also get $\alpha = \frac{\beta}{1-\beta}$ or $\frac{1}{\alpha} - \frac{1}{\beta} = 1$

From this relationship, it is clear that as α approaches unity, β approaches infinity.

The CE configuration is used for almost all transistor application because of its high current gain (β).

Relation among α, β and γ: In the CC configuration, I_B is the input current and I_E is the output current.

$$\therefore \quad \gamma = \frac{\Delta I_E}{\Delta I_B} Z$$

$$\Delta I_B = \Delta I_E - \Delta I_C,$$

we get

$$\therefore \quad \gamma = \frac{\Delta I_E}{\Delta I_E - \Delta I_C}$$

Dividing the numerator and denominator on RHS by ΔI_E, we get

$$\gamma = \frac{\frac{\Delta I_E}{\Delta_E}}{\frac{\Delta_E}{\Delta I_E} - \frac{\Delta I_C}{\Delta I_E}} = \frac{1}{1 - \alpha}$$

$$\boxed{\gamma = \frac{1}{1 - \alpha} = \beta + 1} \tag{3.12}$$

If $\propto = 0.9$, then $\beta = \frac{0.9}{1-0.9} = 9$ and $\gamma = 10$.

3.6   Field-Effect Transistor or UJT

As already discussed in Sect. 3.5 that **a bipolar junction transistor (BJT) is a current-controlled device**, the output characteristics of BJT is controlled by the base current and not the base voltage. Also, the operation of BJT depends on two types of charge carriers, *i.e.* holes and electrons. As bipolar junction transistors (BJT) have the weakness of low input impedance and considerable noise level, therefore, in some of the applications unipolar transistors are used. The operation of unipolar transistors or unijunction transistor (UJT) depends only on one type of charge carrier, *i.e.* either holes or electrons. Such type of transistors is called field-effect transistor (FET) or UJT.

So field-effect transistor (FET) is three-terminal semiconductor device, in which current conduction is by one type of carriers (*i.e.* either holes or electrons) and is controlled by the effect of electric field through the gate voltage. **Therefore, FET is a voltage-controlled device.**

In a broad sense, there are three types of field-effect transistors (shown in Fig. 3.20) as follows:

(i)   Junction field-effect transistor (JFET),
(ii) Metal–oxide–semiconductor field-effect transistor (MOSFET) or insulated-gate field-effect transistor (IGFET) and
(iii) Metal–semiconductor field-effect transistor (MESFET).

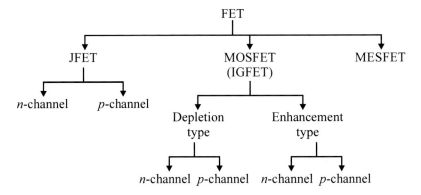

Fig. 3.20 Different types of FET

In metal–semiconductor field-effect transistor (MESFET), the *pn* junction in JFET is replaced by a Schottky barrier diode.

3.6.1   *Junction Field-Effect Transistor (JFET)*

As shown in Fig. 3.20, the junction field-effect transistor (JFET) may be divided into the following two categories depending upon their structure.

A side view of a simplified version of an *n*-channel JFET is shown in Fig. 3.21. JFET consists of an *n*-type semiconductor (silicon) region called channel with two islands of heavily doped p^+-type semiconductor in the sides, thus forming two *pn* junctions. The space between the junction is known as channel. These two p^+ regions are connected with each other (externally or internally) and are called **gate (G)** (Fig. 3.2a). *In fact there is only one gate on one side only in practical JFET as explained in Art. 3.6.2. Two gates are shown only for understanding the working better.* Ohmic contacts are made at the two ends of the *n*-type semiconductor region called channel. One terminal is known as the **source (S)** through which majority carries (electrons in this case) enter the channel. The other terminals are known as **drain (D)** through which these majority carriers leave the channel. Thus, a JFET has essentially three terminals called gate (G), source (*S*) and drain (*D*). Unlike the BJT, the JFET is a unipolar device in which the current is due to majority carries only. Interchanging the *n*- and *p*-type semiconductor results in a *p*-channel JFET. Figure 3.21b shows a side view of a simplified version of a *p*-channel JFET. In this device, holes are majority carriers.

The schematic symbols of *n*-channel JFET and *p*-channel JFET are also shown in Fig. 3.21a and 3.21b, respectively. It may be noted that in an *n*-channel, JFET, the direction of the gate arrow is towards the vertical line. The vertical line represents the *n*-channel; on the other hand, in a *p*-channel JFET symbol, the arrow points away from the vertical line. In this case, the vertical line represents the *p*-channel.

Fig. 3.21 Junction field-effect transistor (JFET) types and their symbols **a** *n*-channel JFET, **b** *p*-channel JFET

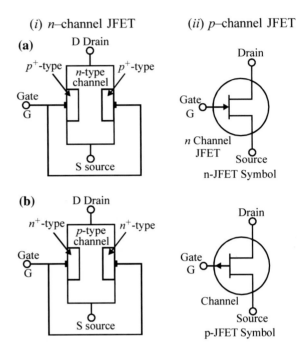

3.6.2 JFET Biasing and Operation

The *n*-channel JFET and *p*-channel JFET biasing are shown in Fig. 3.22a and 3.22b, respectively. In each case, the voltage between gate and source (V_{GG}) is such that the gate is reverse biased. The width of the depletion layer depends upon this reverse biasing, which restricts the flow of majority carriers through the channel.

A high voltage supply (V_{DD}) is connected between the drain and source, which sets up a flow of majority carriers through the channel from source to drain.

'*The drain and source terminals are interchangeable*', *i.e.* either can be used as source and the other end as drain. However, once the JFET is connected to the circuit, the terminals are fixed, *i.e.* the terminal through which the majority carriers enter the channel is called source and the other through which the majority carriers leave the channel is called '*drain*'. Please note the signal phase reversal in Fig. 3.22b (as in BJT: CE and CC configurations).

Operation of JFET

To understand the operational behaviour of a JFET, consider the actual and simplified structures of an *n*-channel JFET as shown in Fig. 3.23a and 3.23b, respectively. **The gate junction is always reverse biased, resulting in a depletion region, which increases with gate reverse voltage**. This depletion region being devoid of majority carriers pinches into the conducting portions of the channel, and hence reduces the drain–source current. Further increase of negative gate voltage

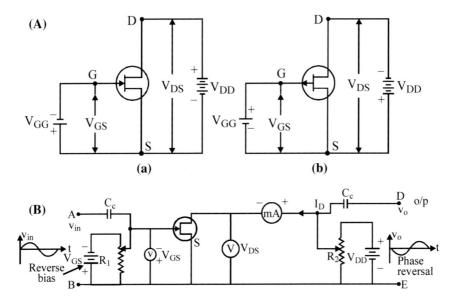

Fig. 3.22 (**A**) JFET polarities and biasing (**a**) *n*-channel (**b**) *p*-channel. (**B**) Circuit arrangement for *n*-channel JFET in common source (CS) mode, along with bias and *ac* signal

will spread the depletion layer further and fully pinch the conducting channel path for drain current (I_D) as shown in details in Figs. 3.23 and 3.24.

The depletion region electric field created by V_{GS} controls the drain current (I_D), that is how the name is field-effect transistor (FET).

The transfer and drain characteristics of JFET are shown in Fig. 3.25a and 3.25b, respectively, for an *n*-channel JFET connected in the common source mode (CS) shown in Fig. 3.23b. Here, the potentiometers R_1 and R_2 are used to vary the voltages V_{GS} and V_{DS}, respectively. The voltages V_{DS} and V_{GS} can be measured by the voltmeters connected across the JFET terminals. The drain current (I_D) can be measured by the milliammeter (mA) connected in series with the JFET and the supply voltage (V_{DD}).

Thus the following points are worth nothing from the characteristics:

(i)   At the initial stage, the drain current (I_D) increases rapidly with the increase in drain–source voltage (V_{DS}) but then becomes almost constant.
     The drain–source voltage (V_{DS}) above which the drain current (I_D) becomes almost constant is called pinch-off voltage.
     The channel is pinched off when

$$V_{DS} = V_{GS} - V_P$$

The pinch off voltage (V_D) reduces with increasing −ve gate bias V_{GS} (see Fig. 3.25b: P_1, P_2, P_3, P_4.)

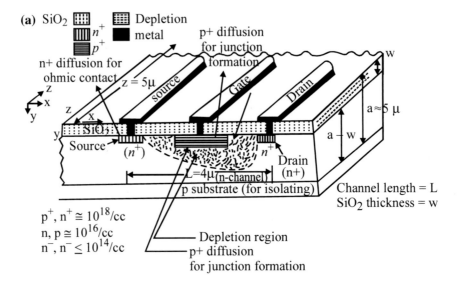

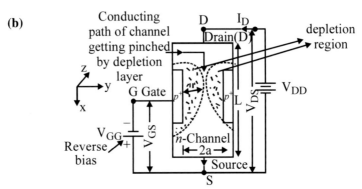

Fig. 3.23 **a** *n*-channel JFET actual layout in planar technology giving typical diffusion density and the measurement of its size, **b** simplified figure just for explaining the working of the *n*-channel JFET. Here, diffusion (gate) is shown on both the sides, which is not actual, but is in one side only

(ii) Due to finite conductivity and resistivity of *n*-channel, there is some voltage drop in V_{DS} as the current flows from terminal *D* to *S*.

(iii) Say at point A in Fig. 3.26a; the p^+n junction is reverse biased by maximum voltage, so that width of the depletion region is extended entirely in the n region near A, where pinch-off takes place.

(iv) As the current flows, the net voltage (potential) at point *B* is less than that at point A. Therefore, the p^+n junction is reverse biased by smaller magnitude and therefore width of depletion region is non-uniform and smaller at point *B* with respect to point *A* (Fig. 3.26a).

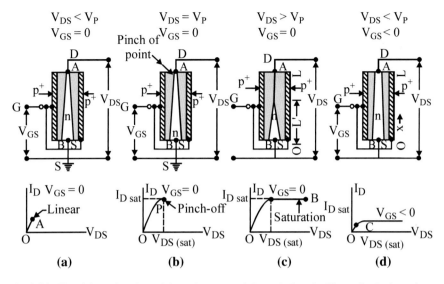

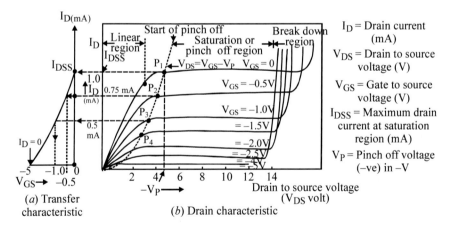

Fig. 3.24 Pictorial explanation of how increase of I_D and the pinching off of channel are connected in *n*-channel JFET, due to depletion region extension. **a** Linear regional point A, **b** Pinch-off point P, **c** Saturation—Point B, **d** Negative gate voltage with very low V_{DS} at point C

Fig. 3.25 **a** Transfer characteristics, **b** Drain characteristics for *n*-channels JFET in Common Source (CS) mode P1, P2, P3, P4 are the starting point of pinch-off on those curves for increasing −ve bias ($-V_{GS}$) from 0 V to −5 V

(v) Therefore, as V_{DS} increases, the depletion layer's width (d) increases so that effective width of the channel decreases (Fig. 3.24c).

(vi) Therefore, when V_{DS} is increased beyond pinch-off also, then the drain current (I_D) first increases and then becomes constant saturation due to further decreased width of channel.

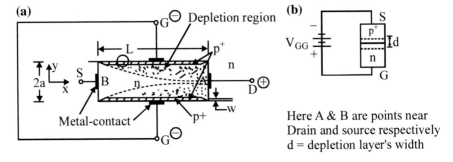

(a) ... Depletion region

(b)

Here A & B are points near Drain and source respectively
d = depletion layer's width

Fig. 3.26 Channelling mechanism in JFET

(vii) The **pinch-off voltage** is the reverse gate voltage that removes all the free charges from the channel (*i.e.* extends the depletion region leaving little region for conduction). Thereafter, the channel current saturates as shown in Figs. 3.24c and 3.25.

The Poisson's equation for the voltage in the *n*-channel in terms of the volume charge density (ρ) is given by (Fig. 3.26):

$$\frac{d^2V(x)}{dy^2} = -\frac{\rho}{\varepsilon_S} = -\frac{q.N_d}{\varepsilon_r\varepsilon_0} \tag{3.13}$$

with N_d = electron concentration density (donor) in the *n*-channel.
$\varepsilon_S, \varepsilon_r, \varepsilon_0$ = the permittivity of material, space and dielectric constant, respectively.

Integrating Eq. 3.13 once and using the boundary condition of electric field $E = \frac{dV(x)}{dy} = 0$ at $y = a$ (the channel width), we get

$$\frac{dV(x)}{dy} = -\frac{q.N_d}{\varepsilon_S}(y-a) \text{ volts/metre} \tag{3.14}$$

Integrating again with the boundary condition $V = 0$ at $y = 0$, we get

$$V(x) = -\frac{q.N_d}{2\varepsilon_S}(y^2 - 2ay) \tag{3.15}$$

The pinch-off voltage $V(x) = V_P$ is at $y = a$; therefore, Eq. 3.15 gives

$$V_P = q.N_d.a^2/(2\varepsilon_S) \tag{3.16}$$

Thus, we see that the pinch-off voltage is a function of doping concentration N_d and channel width 'a'. With fully pinched-off condition, the FET is said to be in the OFF state, with I_d saturated.

As the drain voltage is increased, the Avalanche breakdown across the gate junction takes place, increasing I_d sharply (Fig. 3.25).

Channel resistance can be expressed as

$$R = \frac{\rho L}{A} = \frac{L}{\sigma A} = \frac{L}{2 \cdot \mu_n q N_d z (a - w)} \tag{3.17}$$

μ_n electron mobility,
L distance between source and drain (see Fig. 3.23a),
z length of the channel in z-direction, *i.e.* seen from top surface,
a width of the channel,
w diffusion depth of ohmic-n^+,
ε_s ($\varepsilon_0 \varepsilon_r$) = permittivity of material Si,
A Channel cross section = $z.(a - w)$,
σ Conductivity of channel region and
N_d Doping donor density/cc of the channel.

∴ Using Eqs. 3.16 and 3.17, we can get drain current at pinch-off (*i.e.* saturation current) (as $a \gg w$, therefore $(a - w) \cong a$) as

$$I_{DSS} = I_{DP} = \frac{V_P}{R} = \frac{\mu_n q^2 N_a^2 z a^3}{L \varepsilon_s} \tag{3.18}$$

It has been found experimentally that general equation of drain current (**Shockley equation**) is

$$\boxed{I_D = I_{DSS} \left(1 - V_{GS}/V_p\right)^2} \tag{3.19}$$

Cut-off frequently (*i.e.* highest frequency possible) is

$$f_c = \frac{2 \mu_n q N_d \cdot a^2}{\pi \varepsilon_s L^2} \tag{3.20}$$

and transconductance as

$$g_m = \frac{\partial I_D}{\partial V_{GS}} = \frac{\partial}{\partial V_{GS}} \left[I_{DSS} (1 - V_{GS}/V_P)^2 \right]$$

at V_{DS} constant.

$$g_m = I_{DSS} \times 2 \left[1 - \frac{V_{GS}}{V_P}\right] \left[-\frac{1}{V_P}\right] \tag{3.21}$$

$$g_m = -\frac{2 I_{DSS}}{V_P} \left[1 - \frac{V_{GS}}{V_P}\right]$$

$$g_m = -2 \frac{\sqrt{I_D I_{DSS}}}{V_P} \text{ (By Eq. 3.18)} \tag{3.22}$$

If we get $g_{mo} = g_m = -\frac{2 I_{DSS}}{V_P}$, when $V_{GS} = 0 \text{ V}$

then Eqs. (3.21) and (3.22) can also be written as

$$g_m = g_{mo} \left[1 - \frac{V_{GS}}{V_P} \right] = \frac{-2\sqrt{I_D I_{DSS}}}{V_P} \tag{3.23}$$

3.6.3 JFET Parameters

The following are the main parameters of a JFET:

(i) **ac drain resistance (r_d)**: *ac* drain resistance of a JFET (r_d) is similar to the *ac* plate resistance (r_p) of a vacuum tube used 40 yrs back. It may be defined as the ratio of change in drain–source voltage (ΔV_{DS}) to the change in drain current (ΔI_D) at constant gate–source voltage (V_{GS}).
i.e. ac drain resistance,

$$r_d = \frac{\Delta V_{DS}}{\Delta I_D} \text{ at constant } V_{GS} \tag{3.24}$$

Referring to the JFET characteristic shown in Fig. 3.25b, it is clear that beyond the pinch-off voltage, the change in drain current is very small for a change in drain–source voltage, because the curve is almost flat. Therefore, the value of drain resistance (r_d) is very large ranging from 100 KΩ to hundred megaohms.

For instance, if a change in drain current of 0.01 mA is produced by a change in drain–source voltage of 4 V, then

$$r_d = \frac{4 \text{ V}}{0.01 \text{ mA}} = 400 \text{ K}\Omega$$

Typical value of r_d is about 400 KΩ.

(ii) **Transconductance (g_m)**: Corresponding to the transconductance of the vacuum tube, JFET also has transconductance which determines the control that the gate–source voltage (V_{GS}) has over the drain current (I_D). The prefix 'trans' in the terminology applied to g_m reveals that it establishes a relationship between an output and input qualities. The root word conductance was chosen because g_m is determined by a voltage-to-current ratio similar to the ratio that defines the conductance of a resistor ($G = I/V$). It may be defined as the ratio of change in drain current (ΔI_D) to the change in gate–source voltage (ΔV_{GS}) at constant drain–source voltage (V_{DS}), *i.e.*

$$\text{Transconductance,} \quad \boxed{g_m = \frac{\Delta I_D}{\Delta V_{GS}}} \quad \text{at constant } V_{DS}. \quad (3.25)$$

Transconductance is usually expressed either in mA/volt or microohms. For instance, if a change in gate voltage of 0.1 V causes a change in drain current of 0.35 mA, then

$$g_m = \frac{0.35 \text{ mA}}{0.1 \text{ V}} = 3.5 \text{ mA/V}$$
$$= 3500 \, \mu \text{ mhos}$$

(iii) **Amplification factor (μ):** The amplification factor (μ) of JFET may be defined as follows:
The ratio of change in drain–source voltage (ΔV_{DS}) to the change in gate–source voltage (ΔV_{GS}) at constant drain current (I_D), i.e.

$$\text{Amplification factor,} \quad \mu = \frac{\Delta V_{DS}}{\Delta V_{GS}} \quad \text{at constant } I_D \quad (3.26)$$

Multiplying and dividing the RHS by ΔI_D, we get

$$\mu = \frac{\Delta V_{DS}}{\Delta V_{GS}} \times \frac{\Delta I_D}{\Delta I_D}$$
$$= \frac{\Delta V_{DS}}{\Delta I_D} \times \frac{\Delta I_D}{\Delta V_{GS}}$$

$$\boxed{\mu = r_d \times g_m} \quad (3.27)$$

3.6.4 Comparison Between FET and BJT

Table 3.4 shows the comparison between field-effect transistor (FET) and bipolar junction transistor (BJT).

3.6.5 Advantage and Disadvantage of JFET

A JFET is a voltage-controlled device (similar to a vacuum tube pentode) in which the output current (drain current) is controlled by the input (gate) voltage; therefore, it has the following important *advantages/disadvantages*:

Table 3.4 Comparison of FET (UJT) and BJT

S. no.	Field-effect transistor (FET)	Bipolar junction transistor (BJT)
1.	It is an unipolar device, *i.e.* current in the device is carried either by electrons or holes. The vacuum tube also used to be unipolar (only electron) device	The bipolar device, *i.e.* current in the device is carried by both electrons and holes
2.	It is a voltage-controlled device, *i.e.* voltage at the gate (or drain) terminal controls the amount of current flowing through the device	It is a current-controlled device, *i.e.* the base current controls the amount of collector current
3.	Its input resistance is very high and is of the order of several megaohms (+)	Its input resistance is very low as compared to FET and is of the order of few kilo-ohms (see example 3.1)
4.	It has a negative temperature coefficient at high current levels. It means that current decreases as the temperature increases. This characteristics prevents the FET from thermal break down, *i.e.* has thermal stability (+)	It has a positive temperature coefficient at high current level due to dependance of minority carrier of I_B on temperature. It means that collector current increases with the increase in temperature. This characteristic leads the BJT to thermal breakdown, and therefore has less thermal stability
5.	It does not suffer from minority-carrier storage effects and therefore has higher switching speeds and higher cut-off frequencies (+)	It suffers from minority carrier storage effects, and therefore has lower switching speed and lower cut-off frequencies than that of FET's
6.	It is less noisy than a BJT as the old vacuum tube and is thus more suitable as an input amplifier for low-level signals. It is used extensively in high fidelity frequency-modulated receivers (+)	It comparatively more noisy than a field-effect transistor, due to charge recombination at the base region
7.	It is much simpler to fabricate as an integrated circuit (I_C) and occupies a less space on I_C chip than that of BJT (+)	It is comparatively difficult to fabricate as an integrated circuit (I_C) and occupies more space on I_C chip than that of FET
8.	JFET can tolerate a much higher level of neutron radiation, since they do not depend on minority carriers for their operation (+)	The performance of BJT is degraded by neutron radiation because of the reduction in minority carrier lifetime by radiations
9.	JFET amplifiers have low gain bandwidth product due to the junction capacitive effect and produce more signal distortion except for small-signal analysis. Voltage gain is ≈ 20 or so	BJT amplifiers usually have high gain bandwidth product for CE and CC as their current gain is very high for both small-signal and large-signal analyses ($\beta \approx 100$). The only exception in BJT is CB, where the current gain is extremely low (+)
10.	Transfer characteristic (*i.e.* V_{GS}, I_D curve) is nonlinear	Transfer characteristic (*i.e.* I_b, I_C curve) is closer to linear (+)
11.	Cost is more	Cost is less (+)
12.	*dc* power consumption is 10 times lesser than in BJT (*i.e.* <10 mW or so) (+)	*dc* power consumption is higher than FET (*i.e.* $\approx$100 mW or so)
13.	Very high life longevity (+)	Life longevity is less
14.	Size is smaller than BJT (+)	Size is bigger than FET

Note: The +sign is given where that device is better over the other

Advantages

(i)   JFET has a very high input impedance (of the order of 100 megaohms) which shows a high degree of isolation between the input and output circuits.
(ii) The operation of JFET depends upon the majority carriers (*i.e.* electron in *n*-channel and holes in *p*-channel JFET) which do not cross junctions but moves parallel to it. Therefore, the inherent noise of vacuum tubes (because of high-temperature operation) and those of ordinary transistors (because of junctions) are not present in a JFET.
(iii) In JFET, the risk of thermal runway is avoided since it has a negative temperature coefficient of resistance.
(iv) A JFET has a smaller size, longer life and high efficiency.
(v)   The power gain of JFET is very high, which eliminates the necessity of using driver stage while using it as a power amplifier.

Disadvantages

(i)   Since JFET has high input impedance, the gate voltage has less control over the drain current. Therefore, JFET amplifier has much lower voltage gain than a bipolar amplifier.
(ii) It has low gain bandwidth product.
(iii) It cannot be used as a current amplifier.
(iv) The value of transconductance (g_m) is small.
(v)   Special handling precautions are required to be taken, as a charge surge can cause breakdown in FET.

3.6.6 Applications of JFET

(i)   JFET can be used as an amplifier.
(ii) JFET can be used as a switch.
(iii) It can be used as analog switch in circuits like sample and hold amplitude modulation, ADC/DAC (analog-to-digital or digital-to-analog converters).
(iv) As voltage variable resistor (VVR).
(v)   In digital circuits.
(vi) With JFET, one can make 'buffer amplifier', as a low noise amplifier and phase shift oscillator.

3.7 Metal–Oxide–Semiconductor Field-Effect Transistor (MOSFET)

A metal–oxide–semiconductor field-effect transistor (MOSFET) is a three-terminal device. Like JFET, it has a source, gate and drain. However, unlike JFET, the gate of a MOSFET is insulated from the channel and therefore sometimes it is also called as insulated-gate field-effect transistor (IGFET). Depending upon the action of gate potential, MOSFETs can be classified into two forms.

(i) Enhancement MOSFET
(ii) Depletion MOSFET.

Principle: By applying a transverse electric field across an insulator, deposited on the semiconducting material, the thickness and hence the resistance of a conducting channel of a semiconducting material can be controlled.

In a depletion MOSFET, the controlling electric field reduces the number of majority carriers available for conduction, whereas in the enhancement MOSFET, application of electric field causes an increase in the majority carrier density, *i.e.* enhances it in the conducting regions of the transistor.

Theses two E-MOSFET and D-MOSFET differ in

(i)   Design fabrication
(ii) I_D functioning (See Art 3.7.5)
(iii) I_D symbol

3.7.1 Enhancement Type MOSFET (E-MOSFET)

Figure 3.27a shows three-dimensional schematic structure of *n*-channel E-MOSFET. The circuit connection with this device is shown in Fig. 3.27b along with schematic structure. The circuit symbol of *n*-channel MOSFET is shown in Fig. 3.27c. As there is no continuous channel in an enhancement MOSFET, this condition is represented by the broken line for the channels in the symbols as shown in Fig. 3.27c.

It consists of two highly doped n^+ blocks diffused in a lightly doped *p*-type *Si* substrate. One n^+ region acts as *source S* and the other region acts as *drain D*. These two regions are separated by 1 mil (0.001 inch) or so. The source and drain regions are joined by *n*-channel formed by induced -ve charge due to V_{GS} = +ve which is also called **inversion layer** in the *p*-material as shown. When a voltage is applied between drain and source making drain positive leading to flow of electrons from source to the drain through this *n*-channel. The surface of the device is insulated by means of SiO_2 layer, and metal contact holes are done by etching the SiO_2 by HF, for making metal connections to source and drain. For this, a thin layer of metal aluminium is deposited over the SiO_2 layer. This acts as the gate G (see Fig. 3.27b).

Operation: If the substrate is grounded and a positive voltage is applied to gate, the positive charge on *G* induces an equal negative charge on the substrate side between the source and drain regions. Due to this, an electric field is produced between the source and drain which is directed perpendicular to the plates of the capacitor through the oxide. The negative charge of electrons which are minority carriers in the *p*-type substrate forms an *inversion layer* with carrier density equal to original *p*-density of *p*-substrate. The gate voltage required for this to happen is called the **gate–source threshold voltage** and is designated by V_t. As V_{GS} is increased beyond V_t, the inversion layer acting as channel gets widened or

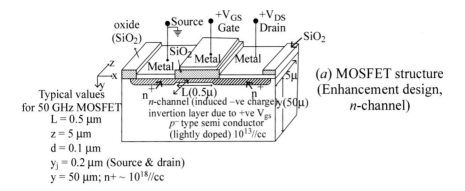

(a) MOSFET structure
(Enhancement design,
n-channel)

Typical values
for 50 GHz MOSFET
L = 0.5 μm
z = 5 μm
d = 0.1 μm
y_j = 0.2 μm (Source & drain)
y = 50 μm; n+ ~ 10^{18}//cc

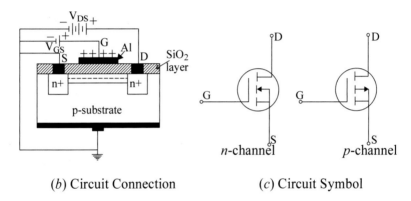

(b) Circuit Connection   (c) Circuit Symbol

Fig. 3.27 n-channel enhancement MOSFET (E-MOSFET) **a** three-channel schematic diagram of MOSFET, **b** circuit connection of n-channel enhancement MOSFET, **c** Enhancement MOSFET circuit symbol, with broken line for channel

enhanced. Thus, once a channel has been established (*i.e.* for $V_{GS} > V_t$), the behaviour of such a device is called an **enhancement MOSFET**. The drain characteristics of a typical n-channel enhancement MOSFET are shown in Fig. 3.28.

Suppose that a channel has been established ($V_{GS} > V_t$) in an enhancement MOSFET, because of the symmetry of the device, this channel will be pinched off when $V_{GS} \leq V_t$, Since $V_{GD} = V_{GS} - V_{DS}$, pinch-off occurs when

$$V_{GS} - V_{DS} \leq V_t \Rightarrow V_{DS} \geq V_{GS} - V_t$$

Under this condition, therefore the MOSFET is in the active region. Of course, this means that it is the ohmic region when $V_{DS} < V_{GS} - V_t$. For the drain characteristics shown in Fig. 3.28, the dashed curve is described by

$$V_{DS} = V_{GS} - V_t \tag{3.28}$$

Therefore, this is the border between the active and the ohmic regions.

When an enhancement MOSFET is in the ohmic region (see Fig. 3.28), it can be shown that up to pinch-off point at saturation of I_D (*i.e.* $I_D = I_{D\,sat}$):

$$I_D = K\left[2(V_{GS} - V_t)V_{DS} - V_{DS}^2\right] \qquad (3.29)$$

Here K is device parameter's constant and if this constant K(ideally) has the same value as for active region operation also, then for small value of V_{DS}, Eq. (3.30) is approximately a linear equation:

$$I_D \approx K\left[2(V_{GS} - V_t)V_{DS}\right] \Rightarrow \frac{I_D}{V_{DS}} \approx 2K(V_{GS} - V_t) \qquad (3.30)$$

Thus for very small values of V_{DS}, the MOSFET behaves as a linear resistor and the approximate value of drain–source resistance r_{DS} is given by (see Fig. 3.28 in lower ohmic region)

$$r_{DS} = \frac{V_{DS}}{I_D} \approx \frac{1}{2K(V_{GS} - V_t)} \qquad (3.31)$$

Now, as it is mentioned earlier that when $V_{DS} = V_{GS} - V_t$ corresponds to the boundary between active and ohmic regions and then substituting this fact in drain current Eq. (3.29), we get the equation of this dashed curve of Fig. 3.28 which gives transfer characteristics (I_D vs V_{Gs}) as parabola, *i.e.* nonlinear in nature and is shown in Fig. 3.29 and in equations as

$$I_{D\,sat} = K(V_{GS} - V_t)^2 = KV_{D\,sat} \qquad (3.32)$$

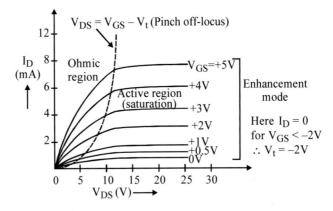

Fig. 3.28 Volt–ampere characteristics of *n*-channel enhancement design MOSFET. Here the '*n*' channel is formed by carrier inversion in the channel region which was originally *p*-type before threshold voltage V_t

Fig. 3.29 Transfer
characteristics of an
enhancement design
MOSFET (E-MOSFET)

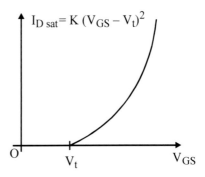

3.7.2 Depletion Type MOSFET (D-MOSFET)

In this design (Fig. 3.30a), we see that $I_d \neq 0$ for $V_{gs} = 0$ with $V_{ds} > 0$, see
Fig. 3.30b.

This is because of n-type of light, doping (10^{15}/cc) is done in the channel region
during fabrication (and hence called soft junction), whereas in (Enhancement)

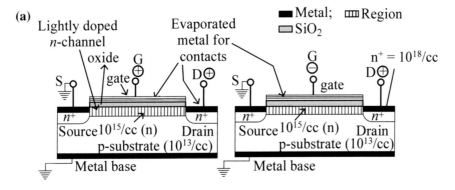

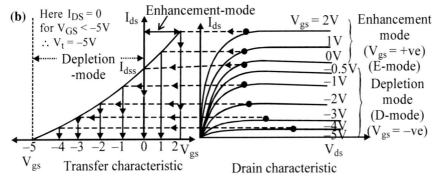

Fig. 3.30 a Depletion design structure of n-channel ON-MOSFET operated in (i) Enhancement
mode (gate +ve), (ii) Depletion mode (gate −ve). **b** Current characteristic of depletion design of
n-channel ON-MOSFET- (or DMOSFET) transfer and drain characteristic, for working in
enhancement and depletion modes

MOSFET, no such doping is there. Therefore, this design can function in the enhancement mode (+ve V_{GS}) as well as in depletion mode (−ve V_{gs}) as well as in depletion mode (−ve V_{GS}) as the doped n-channel already exists even with $V_{gs} = 0$. ∴ Two modes are

(a) **The +ve gate voltage** will induce/enhance more −ve carrier in this n-channel (like a capacitor effect due to SiO_2 insulator) and hence called enhancement mode (Fig. 3.30a(i) and 3.30b) operation.
(b) **The −ve gate voltage** will induce the +ve carries in the channel, thereby neutralizing/reducing/depleting the −ve carriers of the channel (Fig. 3.30a(ii) and 3.30b), and hence called **depletion mode**.

Thus, we see that the enhancement design MOSFET (E-MOSFET) can function in enhancement mode only, while the depletion design MOSFET (DMOSFET) can work in depletion mode as well as enhancement mode.

In the enhancement design type, the increase in V_{ds} will not increase the I_{ds} as there is no n-channel for conduction; therefore, this design type is also called **OFF-MOSFET**. In case of depletion design type, the I_{ds} can increase with V_{ds} even with $V_{gs} = 0$, as the channel carriers are present (due to doped layer) and therefore this design type is also called **ON-MOSFET**.

So far we have discussed the n-channel MOSFET only, but all these are true for p-channel also, with n replaced by p and biasing sign reversed.

Applications

Both of the MOSFET are generally used as power amplifier, as they have some advantages over BJT, JFET and MESFET, e.g.

1. It can be linear power amplifier in the enhancement mode as C_{in} and g_m does not depend on V_g, while C_{out} is independent of V_{ds}. E-MOSFET is used as invertor and active load in digital circuits.
2. Gate *ac* input signal can be quite large as n-channel depletion-type DMOSFET can operate from depletion mode region (−V_g) to enhancement mode region (+V_g). D-MOSFET is used as a switch also.

Advantages: It has all the advantages of JFET (Page 139 Art. 3.6.5).

Gate Protection in MOSFET

The SiO_2 layer of the gate being extremely thin gets damaged readily on application of large voltages caused by charge. An open-circuited gate may accumulate enough charge so as to produce an electric field large enough to puncture the dielectric. To avoid the possibility of damage through this cause, MOS device may be fabricated with a Zener diode placed between the gate and substrate. Under normal operating conditions, this Zener diode is open and it does not influence the working of the circuit. However, in case the voltage at the gate becomes excessive, then the Zener diode breaks down (normally without getting damaged), so that the gate potential gets limited to a maximum value, equal to the Zener breakdown voltage.

Table 3.5 Comparison of JFET and MOSFET

S. No.	JFET	MOSFET (Both types)
1.	JFET are two types p- and n-channel	MOSFET can be of depletion design type or enhancement design type
2.	**Symbols** n-channel JFET p-channel JFET	 n-channel depletion MOSFET n-channel enhancement MOSFET (For p-channel, the only difference is the arrow reversal)
3.	JFETs do not have the insulated gate	MOSFETs have insulated gate due to SiO_2 layer
4.	Input impedance of JFET is lower than that of the MOSFET (Ri $\approx$ 500 KΩ) ; C_i = v.low	Input impedance is higher due to the insulated gate structure (Ri > 1 mΩ); C_i = high due to SiO_2
5.	Drain resistance is lower than that of MOSFET ($R_D \approx$ Few KΩ)	Drain resistance is higher than that of JFET ($R_D \approx$ 10's of KΩ)
6.	Feedback resistance are very much smaller than that of MOSFET	Feedback resistance are much larger than that of JFET
7.	Biasing used are self-bias, fixed and simple voltage divider	Biasing of depletion MOSFET are similar to JFET but circuits for enhancement MOSFET are drain feedback biasing and simple voltage divider type

3.7.3   Comparison of JFET and MOSFETs

See Table 3.5.

3.7.4   Comparison of JFET and D-MOSFET

In this section, let us compare JFET and DMOSFET in the form of Table 3.6.

Examples Related to BJT and FET

Example 3.2 A transistor has an emitter current of 10 mA and a collector current 9.95 mA. Calculate its base current.

(**U.P. Technical University Special Exam 2001**)

Table 3.6 Comparing JFET and DMOSFET

S. no.	JFET	DMOSFET
1.	Symbol of *n*-channel JFET	Symbol of *n*-channel depletion-type MOSFET
2.	The gate is not isolated from the channel	The layer of SiO_2 is present between the gate and channel as isolator
3.	Reduction in I_D is due to the narrowing of the channel width with more negative V_{GS}	Reduction in I_D is due to the hole-electron recombination process taking place under the influence of negative V_{GS}
4.	Drain characteristics are drawn only for $V_{GS} \leq 0$ volts	Drain characteristics are same as those of JFET except for the part corresponding to positive V_{GS}
5.	Input impedance is high (over 500 KΩ)	Input impedance is still higher due to the presence of SiO_2 layer (over 1 mΩ)
6.	Voltage gain is around 20–30	Voltage gain is <10

Solution Given $I_E = 10$ mA, $I_C = 9.95$ mA

$$I_E = I_C + I_B$$
$$I_B = I_E - I_C = 0.05 \, \text{mA}$$

Example 3.3 Figure 3.31 shows *npn* transistor for which $I_E = 10$ mA and $\beta = 100$. Find the value of I_C and I_B (**M.D. University, 2006**).

Fig. 3.31 *npn* transistor of Example 3.3

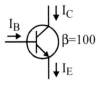

Solution We know that

$$\alpha = \beta/(1+\beta)$$

$$\alpha = \frac{I_C}{I_E}$$

$$\beta = \frac{I_C}{I_B}$$

$$\alpha = \frac{100}{100+1} = 0.99$$

$$I_C = \alpha I_E = 0.99 \times 10 = 9.9\,\text{mA}$$

$$I_B = I_E - I_C = 10 - 9.9 = 0.1\,\text{mA} \quad \textbf{Ans.}$$

Example 3.4 For an *npn* transistor $\propto\, = 0.995$, $I_E = 10$ mA, leakage current I_{CBO} (For I_{CO}) $= 0.5$ mA. Determine I_C, I_B, β and I_{CEO}.

Solution

$$I_C = \alpha I_E + I_{CBO} = 0.995 \times 10\,\text{mA} + 0.5\,\mu\text{A}$$
$$= (9.95 + 0.0005)\,\text{mA}$$
$$= 9.9505\,\text{mA}$$

$$\therefore \quad I_B = I_E - I_C = (10 - 9.9505)\,\text{mA} = 0.0495\,\text{mA} = 49.5\,\mu\text{A}$$

$$\beta = \frac{\alpha}{1-\alpha} = \frac{0.995}{1-0.995} = 199$$

$$I_C = \beta \cdot I_B + I_{CEO}$$

$$\therefore \quad I_{CEO} = I_C - \beta \cdot I_B$$
$$= 9.9505\,\text{mA} - 199 \times 0.0495\,\text{mA}$$
$$= 0.1\,\text{mA} = 100\,\mu\text{A} \quad \textbf{Ans.}$$

Example 3.5 The following information is included on the data sheet of an *n*-channel JFET.

$$I_{DSS} = 20\,\text{mA}; \quad V_P = -8\,\text{Volt}$$
$$g_{mo} = 5000\,\mu\text{mhos}$$

and,

Find the values of the drain current and transconductor at $V_{GS} = -4$ V.

Solution Given that

$$I_{DSS} = 20 \text{ mA}$$
and $V_P = -8 \text{ V};$ $\quad g_{mo} = 5000 \, \mu\text{mhos}$
$$V_{GS} = -4 \text{ V}$$

we know that

$$I_D = I_{DSS} \left(1 - \frac{V_{GS}}{V_P} \right)^2$$

$$= 20 \left(1 - \frac{(-4)}{(-8)} \right)^2 = 5 \text{ mA}$$

Also, transconductor is given by

$$g_m = g m_0 \left(1 - \frac{V_{GS}}{V_P} \right) = 5000 \left(1 - \frac{(-4)}{(-8)} \right) = 2500 \, \mu \, \text{mhos} \quad \textbf{Ans.}$$

Example 3.6 The data sheet for a certain enhancement-type MOSFET reveals that $I_{D(on)} = 10 \text{ mA}$ at $V_{GS} = -12 \text{ V}$ and $V_{GS(th)} = -3 \text{ V}$. Is this device p-channel or n-channel? Find the value of I_D, when $V_{GS} = -6 \text{ V}$.

Solution Given that $I_{D(on)} = 10 \text{ mA}; V_{GS} = -12 \text{ V}$ and $V_{GS(th)} = -3 \text{ V}$.
Since the value of V_{GS} is negative for the enhancement-type MOSFET, this indicated that the device is p-channel. Also $V_{GS(th)} = V_t$.

We know that the saturated drain current is given by

$$I_D = \text{K} \left[V_{GS} - V_{GS(th)} \right]^2$$
or $I_{D(on)} = \text{K} \left[V_{GS} - V_{GS(th)} \right]^2$
$$10 = K[-12 - (-3)]^2 = 81 \text{ K}$$
$$K = 10/81 = 0.12 \text{ mA/V}$$

Substituting this value of K and V_{GS} (equal to -6) in the equation of I_D above,

$$I_D = 0.12 \left[-6 - (-3) \right]^2 = 1.08 \text{ mA} \quad \textbf{Ans.}$$

Example 3.7 Show that if $|V_{GS}| \ll |V_P|$ then drain current can be approximated as

$$I_{DS} = I_{DSS} - g_{mo} V_{GS}$$

Solution We know that

$$I_{DS} = I_{DSS} \left[1 - \frac{V_{GS}}{V_P} \right]^2$$

Simplifying, we get $I_{DS} = I_{DSS}\left[1 - 2\frac{V_{GS}}{V_P} + \left(\frac{V_{GS}}{V_P}\right)^2\right]^2$

Neglecting the square term, we get

$$I_{DS} = I_{DSS} - \frac{2I_{DSS}V_{GS}}{V_P}$$

Putting $\qquad g_{mo} = -\dfrac{2I_{DSS}V_{GS}}{V_P}$

$$I_{DS} = I_{DSS} - g_{mo}V_{GS} \quad \textbf{Ans.}$$

Example 3.8 An n-channel JFET has $I_{DSS} = 8$ mA and $V_P = -5$ V. Find the minimum value of V_{DS} for pinch-off region and the drain current I_{DS} for $V_{GS} = -2$ V in the pinch-off region.

Solution The minimum value of V_{DS} for pinch-off to occur for $V_{GS} = -2$ V is

$$V_{DS(min)} = V_{GS} - V_P = -2 - (-5) = 3 \text{ V}$$

$$I_{DS} = I_{DSS}\left[1 - \frac{V_{GS}}{V_P}\right]^2$$

Substituting values, we get

$$I_{DS} = 8 \times 10^{-3}[1 - (-2)/(-5)]^2$$

Solving, we get $I_{DS} = 2.88$ mA

Example 3.9 $V_P = -2$ V, $I_{DSS} = 1.65$ mA for the circuit in Fig. 3.32. It is desired to bias the circuit at $I_D = 0.8$ mA, $V_{DD} = 24$ V. Find (i) V_{GS} and (ii) g_m (Fig. 3.31).

Solution

(i)   As we know that

Fig. 3.32 n-channel JFET of Example 3.9

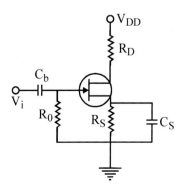

$$V_{GS} = V_P \left[1 - \sqrt{\frac{I_D}{I_{DSS}}} \right]$$

$$= -2 \left[1 - \sqrt{\frac{0.8 \times 10^{-3}}{1.65 \times 10^{-3}}} \right]$$

$$V_{GS} = -0.6074 \text{ V}$$

$$g_{mo} = \frac{-2I_{DSS}}{V_P}$$

$$= \frac{-2 \times 1.65}{-2} = 1.65 \text{ milli mhos}$$

(ii) $$g_m = g_{mo} \left(1 - \frac{V_{GS}}{V_P} \right)$$

$$= 1.65 \times 10^{-3} \left(1 - \frac{-0.6074}{-2} \right)$$

$$g_m = 1.15 \text{ milli mhos}$$

Example 3.10 The pinch-down voltage of p-channel junction FET is $V_P = 5$ V and the drain-to-source saturation current $I_{DSS} = -40$ mA. The value of drain–source voltage V_{DS} is such that the transistor is operating in the saturated region. The drain current is given as $I_D = 15$ mA. Determine the gate–source voltage V_{GS} (**GATE 2001**).

Solution It is found experimentally that a square-law characteristic closely approximate the drain current in saturation:

$$I_{D(Sat)} = I_{DSS} \left(1 + \frac{V_{GS}}{V_P} \right)^2$$

Given that

$$V_P = 5 \text{V}$$
$$I_{DSS} = -40 \text{ mA}$$
$$I_D = -15 \text{ mA}$$

$$-15 \text{ mA} = -40 \text{ mA} \left(1 + \frac{V_{GS}}{5} \right)^2$$

On solving $0.612 = 1 + \frac{V_{GS}}{5}$

$$V_{GS} = 1.938 \text{ V} \quad \textbf{Ans.}$$

Example 3.11 The following readings were obtained experimentally from a FET:

$$
\begin{array}{llll}
V_{GS} & -0.1\,\text{V} & -0.1\,\text{V} & -0.4\,\text{V} \\
V_{DS} & 5\,\text{V} & 14\,\text{V} & 14\,\text{V} \\
I_D & 8\,\text{mA} & 3.3\,\text{mA} & 7.1\,\text{mA}
\end{array}
$$

Obtain (i) *ac* drain resistance, (ii) transconductance and (iii) amplification factor.

Solution

(i) At constant V_{GS} (*i.e.* at $V_{GS}= 0.1$ V), when V_{DS} is increased from 5 V to 14 V, the drain current I_D increases from 8 to 3.3 mA.

Therefore, change in drain–source voltage, $\Delta V_{DS} = 14 - 5 = 9\,\text{V}$.
Change in drain current, $\Delta I_{DS} = 3.3 - 8 = 0.3$ mA
Hence, *ac* drain resistance will be at constant V_{GS}:

$$
r_d = \frac{\Delta V_{DS}}{\Delta I_D}
$$

∴ At constant V_{Gs}, $r_d = \frac{9\,\text{V}}{0.3\,\text{mA}} = 30\ \text{K}\Omega$.

(ii) At constant V_{DS} (*i.e.* at $V_{DS} = 14$ V), when V_{GS} is changed from -0.1 to -0.4 V the drain current I_D changes from 8.3 to 7.1 mA.

The change in gate–source voltage will be

$$
\Delta V_{GS} = 0.4 - 0.1 = 0.3\ \text{V}
$$

The change in drain current will be

$$
\Delta I_D = 3.3 - 7.1 = 1.2\ \text{mA}
$$

Therefore, transconductance will be

$$
g_m = \frac{\Delta I_D}{\Delta V_{GS}} \text{ at constant } V_{DS}
$$

$$
g_m = \frac{1.2\ \text{mA}}{0.3\ \text{V}} = 4\ \text{mA/V} = 4000\ \mu\text{mhos}
$$

(iii) Amplification factor is given by

$$
\mu = r_d \times g_m = (30 \times 10^3) \times (4000 \times 10^{-6}) = 120
$$

3.7.5　Comparison of E-MOSFET and D-MOSFET

In this section, we will compare the two n-channel -MOSFET's, i.e. Enhancement type and depletion type MOSFET: E-MOSFET and D-MOSFET, which differ in design, functioning and symbol.

Difference	n-channel E-MOSFET	n-channel D-MOSFET
1. Design	As per Fig. 3.27b and 3.30a, we see that in E-MOSFET, there is no diffused channel below SiO_2 between drain and source	As per Fig. 3.30, below SiO_2 layer of gate, lightly n-doped ($\approx 10^{15}$/CC) region is fabricated, making a soft junction with p-substrate as well as a n-channel
2. Functioning of n-channel	The n-channel is induced at the top of the lightly doped ($\approx 10^{13}$/CC) p-substrate by +ve gate bias, across SiO_2 dielectric. Here $V_{gs} = 0$, $I_{ds} = 0$ therefore we require V_{gs} = +ve bias for enhancing/forming the n-channel for current I_{ds} to fow (see Fig. 3.28)	The free fabricated n-channel, even at $V_{gs} = 0$; $I_{ds} \neq 0$ (is a few mA). Therefore lowering charge carriers for low I_{ds} (Depletion mode). We need −ve gate bias and for enhancing I_{ds} (denser charge carriers) (i.e. Enhancement mode), we need +ve gate bias (see Fig. 3.30b)
3. Symbol (for p-channel the arrow is reversed)	 (The dotted channel line shows no pre-diffused n-channel)	 (Solid channel line shows pre-diffused n-channel)

Questions

Fill in the blanks:

1. For a large value of $|V_{DS}|$, a FET behaves as _____.
2. In a JFET, at pinch-off voltage applied on the gate the drain current is almost at _____ value.
3. The majority charge carriers in the emitter of an NPN transistor are _____.
4. The common collector (CC) configuration is used for _____.
5. The common emitter (CE) configuration is used for _____.

6. The common base (CB) configuration is used for _____.
7. The common collector (CC) amplifier is also known as _____.
8. Bipolar junction transistor is a _____ controlled device.
9. Field-effect transistor is a _____ controlled device.
10. Compared to bipolar transistor, a JFET has _____ impedance and low voltage gain.
11. For a JFET, when V_{DS} is increased beyond the pinch-off voltage, the drain current _____.
12. n-channel FETs are superior to p-channel FETs, because mobility of electrons is _____ than that of holes.
13. Field-effect transistor has _____ impedance at input side.
14. In integrated circuits, *npn* construction is preferred to *pnp* construction because _____.
15. All FETs are _____ input _____ amplifier.
16. BJT is _____ controlled device whereas FET is _____ controlled device
17. A BJT has a base current of 250 μA and emitter current of 15 mA. Determine the collector current _____.

Short Questions:

1. Draw the symbol of NPN and PNP transistors and specify the functions of the leads.
2. Draw the input and output characteristics of CE configuration with proper biasing.
3. Draw the input and output characteristics of CB configuration with proper biasing.
4. Draw the input and output characteristics of CC configuration with proper biasing.
5. Draw V-I characteristics of CS JFET. Where do we use the ohmic region?
6. Why is collector wider than emitter and base? Why is the base region made very thin? What are the applications of BJT?
7. What is a JFET? What are the differences between JFET and bipolar transistor?
8. Define pinch-off voltage in JFET and give the structure.
9. Thermal stability of performance is better in FET than BJT. Explain why?
10. What is the relationship among $\propto$, β and γ.
11. What is the relationship among μ, g_m and r_d.
12. What is a MOSFET? What are the difference between JFET and MOSFET?
13. Why Z_{IN} in MOSFET is very high? In what way it is useful in the circuit?
14. 'Minority base current I_B controls I_C in BJT, therefore it has less thermal stability', explain.
15. Explain (a) transistor action, (b) It is possible because of very thin base only.

Long Questions:

1. Explain the working of bipolar junction transistors in active, cut-off and saturation regions.
2. Explain the input and output characteristics of transistors in CB configuration.
3. Explain the input and output characteristics of transistors in CE configuration.
4. Explain the input and output characteristics of transistors in CC configuration.
5. Explain the operation of PNP and NPN transistors.
6. With help of neat sketches, explain the construction and operation of n-channel JFET.
7. With help of neat sketches, explain the construction and operation of MOSFET.
8. Explain about the characteristics of JFET.
9. Explain about the channelling effect in JFET.

Chapter 4
Optoelectric Devices

Contents

4.1   Introduction

Optoelectric (also called photoelectric) devices are devices which use light or electric current as generator of the other or both together for controlling something else/producing another type of effect. These devices can

© Springer Nature Singapore Pte Ltd. 2020
S. S. Srikant and P. K. Chaturvedi, *Basic Electronics Engineering*,
https://doi.org/10.1007/978-981-13-7414-2_4

1. Emit light when given a bias (e.g. LED).
2. Convert light into electricity (e.g. solar cells).
3. Detect light intensity by measure of its parameters which change with light (e.g. LDR, photodiode detector, PIN diode, phototransistor, etc.).
4. Trigger another light-sensitive device by light of ON LED without any wire/line in an isolated way (e.g. optocoupler).
5. Rotate polarized light in the absence of electric field, but not in its presence (e.g. liquid crystal display, *i.e.* LCD).

Knowledge of wavelength, energy gap (E_g) and visible light region with eye sensitivity will be useful as given in Fig. 4.1.

Limits units: There are three units for measuring light from a source, out of them, the most popular is lux, *i.e.* mW/cm^2.

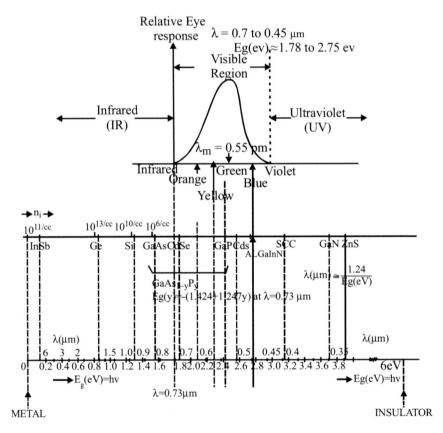

Fig. 4.1a Relative sensitivity of the human eye for different colours. The corresponding photon energy and band gaps for some semiconductors are also shown

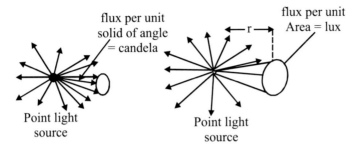

Fig. 4.1b Light intensity in Candela and Lux

(a) **Total light energy (luminous flux ϕ_s):** This is total energy coming out of a source and measured in milliwatt (mV) or lumen (lm),

where 1 lm = 1 lm = 1.496 mW.

(b) **Light Intensity (E_A):** As the luminous flux (ϕ_s) spread over spherical area of $4\pi r^2$, therefore, at distance r, the Lumen density per unit area is called light intensity (E_A).

$$E_A = \phi_s/4\pi r^2 \qquad (4.1)$$

The unit of ϕ_s is lux(lx) and is lumens/metre^2. Light intensity can be in mW/cm^2 or lumen/feet^2 (lm/ft^2), the latter known as foot candle (f_c), where $1f_c = 10.764$ Lx.

(c) **Inverted and bold intensity (luminance (E_S)):** This is luminous flux per unit solid angle (or cone) emitting from a source and is one lumen/solid angle having unit called Candela (C_d).

$$E_S = \phi_s/4\pi \qquad (4.2)$$

Now we will discuss some of the important optoelectric devices e.g. *pn* junction diodes (LED, solar cell, photodetector), transistor photodetectors, optocoupler, LDR, IR emitter and LCD.

4.2 Light-Emitting Diode (LED)

LED is an abbreviation of light-emitting diode. An LED-based display is perhaps the most important in the display devices used for various applications.

Some semiconductor materials used in LED manufacturing and their emitting colour of light is given in Table 4.1. These materials are normally mix of elements of III and V Group of Mendeleev Table for two-element semiconductors or of III, IV, V Group for three-element semiconductor.

Table 4.1 Different colours emitted by LED of different materials

Colour	Wavelength $\lambda(\mu \cdot m) = \frac{1.24}{E_g(eV)}$	Band gap $E_g(eV)$	Approx. V_F at 20 mA	Semiconductor material X = of LED for pn junction
Infrared (IR)	$\lambda > 0.75\ \mu m$	$E_g < 1.74$	1.2 V	GaAs; GaAlAs;Si;Ge; InSb; GaAlP
Red	0.75–0.73 μm	1.74–1.7	1.7–1.8	GaAsP; CdSe; $Al_{0.3}\ Ga_{0.7}As$; $As_{0.6}P_{0.4}$
Orange and yellow	0.73–0.58 μm	1.9–2.1 2.15– 2.39	1.9–2.4	InGaP; GaP, SiC
Green	0.58–0.50 μm	2.4–2.6	2.5–2.9	InGaN; GaN; AlGaP
Blue	0.50–0.45 μm	2.6–2.8	2.6–3.0	ZnSe; InGaN;SiC
Violet	0.45–0.40 μm	2.81–3.6	3.0–3.5	InGaN
Ultraviolet	$\lambda < 0.4$ μm	>3.6	>3.6 V	Above violet material + Yellow phosphorus
White	0.7–0.4 μm	–	≈4.0 V	(Vitrium aluminium garnet) or GaInN + Phosphor coating

Note $GaAs_{1-x}\ P_x$ can give colours from I_R to yellow, depending upon the value to x. At $x = 0$ is GaAs (infrared) while at $x = 1$, we get GaP (yellow). Here for $GaAs_{1-x}\ P_x$; $E_g = (1.424 + 1.247x)$ eV

4.3 Principle of LED

An LED is moderately ($>10^{16}$/cc) doped pn junction made from some special materials like GaAs, GaP, GaN, InP, AlN, etc., which have energy band gap lying between 1.4 eV (infrared) to 3.4 eV (ultraviolet), as shown in Table 4.1.

Radiation emission: When this diode is forward biased, then some of the electrons of the valence band (E_V) absorbs energy/get excited and jump to the conduction band (E_C), creating holes in valence band. At the same time, some of the excited electrons keep returning back to the valence band (E_V) to recombine with the holes they left, thereby losing energy (*i.e.* emitting earlier absorbed energy). In this process of recombination with holes (Fig. 4.2), this energy ($E_g = h\nu$) emitted is equivalent to the optical radiation energy getting emitted having frequency ν.

Recombination may be of two types:

1. Direct recombination → E_C to E_V directly as shown in Fig. 4.2a.
2. Indirect recombination → E_C to E_V after stopping on a trap level (as shown in Fig. 4.2b). This will lead to two types of radiations at frequencies ν_1 and ν_2.

The construction of an LED is shown in Fig. 4.3a in chip form and in Fig. 4.3b in the packaged form Fig. 4.3c shows the LED that can give two colours, when two chips (LEDs) are fixed on a single conductor A, with two wires for the two LED's, biasing. When AK_1 is forward biased, it gives one colour and with AK_2 biasing another colour. Therefore, a.c. signal of low frequency (<10 c/s) between AK_1 and AK_2 will give slow flickering of alternate colour, which will be visible to our eyes.

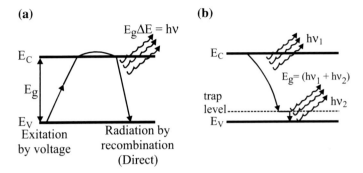

Fig. 4.2 a Direct recombination, **b** indirect recombination

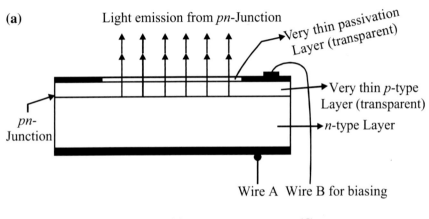

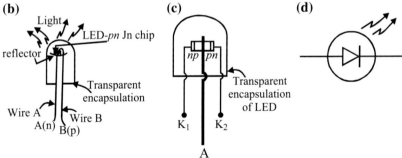

Fig. 4.3 a Combination of an LED semiconductor *pn* junction, **b** The active transparent encapsulated LED, **c** Bi-colour LED and **d** Symbol of LED

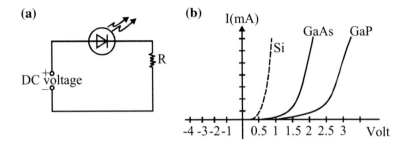

Fig. 4.4 a LED connected with series resistance R showing biasing **b** LED V-I characteristics

Forward biasing of a LED is normally through a series resistance for protecting against heavy current flow, which will damage it, as shown in Fig. 4.4a.

The turn-ON voltage of LED is basically from 1.5 to 5V with the current varying from 10 to 15 mA with power dissipation of 10 mW to 150 mW depending upon the semiconductor used. Being moderately to highly doped (10^{16}–10^{17}/cc), the reverse breakdown voltage is less than 10 V. The properties of LED are as follows:

(a) It can emit different colours with different semiconductor materials.
(b) Intensity is proportional to the forward bias current.
(c) Switching speed of ON to OFF and OFF to ON is as low as 1 n-s.
(d) Life longevity period is as high as 100,000 hrs.

The LEDs are widely used as ON/OFF power indicator light in burglar alarm systems, phone, multimeters, calculators, digital meters, microprocessors, digital computers, electronic telephone exchange, intercom, electronic panels, digital watches, video displays and optical communication systems.

4.4 Photovoltaic Cell or Solar Cell

It is a *pn* junction diode mostly of Si material, where the visible light can be converted directly into electricity. These devices have long being used for the power supplies of satellite, space vehicles as well as the simple calculators. On a sunny day, 1 KW/m² solar power is available from sun. Solar cells have maximum efficiency of 25%, *i.e.* we get 250 W/m² using solar cells. Normal solar cells have only 10% efficiency. The total solar power on earth is 600 TW, whereas installed solar power so far is 1 GW around the world, *i.e.* only $1.6 \times 10^{-4}\%$

When light falls on the *pn* junction through the thin n layer (nearly transparent being very thin 0.5 μm or so) then electron–hole pair (EHP) gets generated at the junction by way of excitation of electron jumping from E_V to E_C (like that in LED) by voltage bias (Fig. 4.2a). Physically the electron gets accumulated, *i.e.* get shifted to the top majority n layer and while the holes to the lower layer of p majority. **This makes them like reservoirs of electrons and holes as shown in Fig. 4.5, with voltage** generated in reverse bias direction from n to p inside diode, see Fig. 4.5.

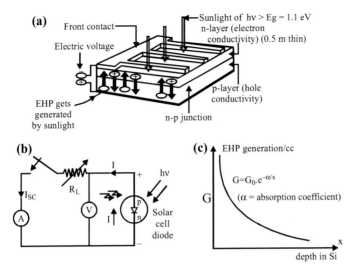

Fig. 4.5 a Electron–hole generation by sunlight and the voltage generated by reservoir of 'e'–'h' charges, in the n and p sides respectively, **b** photovoltaic cell when connected to load (R$_L$), **c** characteristics curve between EHP generation and depth in Si for absorption coefficient of photovoltaic cell

Therefore, when a photovoltaic cell is kept under illumination (for example, under the sun) and is connected across a load, as shown in Fig. 4.5b, then a small current flows. Here the conversion efficiency (light into current) depends on the spectral content and intensity of illumination. To capture and convert more energy (from the sun) into electric energy, the photovoltaic cells are linked to form photovoltaic arrays.

An array is simply a large parallel and series combination of cells connected by wires, for increasing the current and voltage, respectively. Figure 4.6a shows the typical characteristics of photovoltaic cell for open-circuit voltage and for short-circuit current for different intensities of light. The current, voltage as well as power characteristics for illuminations of 100 and 50 mW/cm^2 are shown in Fig. 4.6b. Here V_{OC}, I_{SC} depend on light intensity and doping in the cell. Here I_{SC} falls with E_g while V_{OC} rises. Here V_m, I_m are voltage and current at P_{max}.

Question 4.1: How different semiconductors are used in fabricating photovoltaic cell or solar cell?

Answer: A layer of n-AlGaAs ($E_g \approx 2$ eV) can be grown on lattice-matched material like p-GaAs ($E_g \approx 1.4$ eV). For reducing surface recombination, one may also use silicon photovoltaic cell ($E_g < 1.4$ eV).

Question 4.2: Why a photovoltaic or solar cell operates in the fourth quadrant of the junction I-V characteristics (Fig. 4.8).

Answer: Because fourth quadrant represents −ve power generation with voltage generated opposite to the forward bias direction, Here let us define fill factor of photovoltaic cell, which is the ratio of maximum power generated $P_m = I_m V_m$ which can be delivered to load out of the P_{max}, (which is equal to that of the product V_{OC}. I_{SC} of a photovoltaic (solar) cell). Practically, it is less than that maximum.

Fig. 4.6 a Short-circuit current (I_{SC}) and open-circuit voltage (V_{OC}) (for a typical solar cell) as a function of illumination intensity, **b** current, voltage and power output characteristics for a typical photovoltaic cell, for two different illuminations

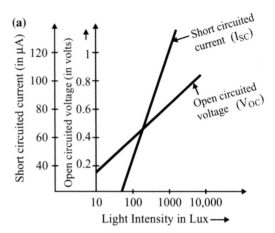

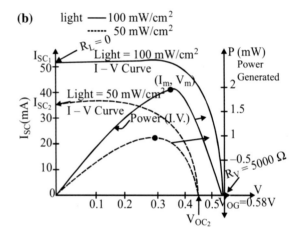

$\therefore$ We define Fill factor $= \frac{V_m \cdot I_m}{V_{OC} \cdot I_{SC}}$ (Usually value is 0.4–0.7).

This parameter works as figure of merit in designing a solar cell. Photovoltaic cells work in the fourth quadrant of the I-V curve as shown in Fig. 4.7a, b because power is generated in solar (or photovoltaic) cell in the quadrant $(-I, +V)$, as $I \cdot V = -ve$. The other devices, e.g. LED, normal diode work in the first quadrant while photodiode detectors $(-I, -V)$ in the third quadrant as shown in Fig. 4.8 where power $(= I\,V)$ is consumed.

Positive properties of photovoltaic devices

1. These transducers are self-generating type, *i.e.* active. They do not require any external power for their operation.
2. Operate over a wide range of temperature, *i.e.* from -100 to $125\,°C$.
3. Offer extremely fast response.

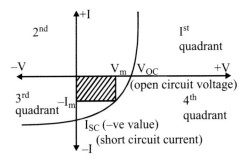

Fig. 4.7 Photovoltaic cell operates in the fourth quadrant of the I-V characteristic

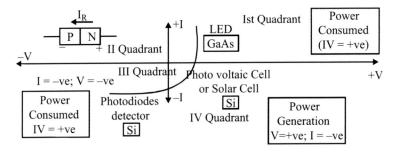

Fig. 4.8 Power is generated (I = −ve, V = +ve, V × I = −ve) in fourth quadrant in devices like photovoltaic cell, while power is consumed (V × I = +ve) in the first quadrant in LED and also in the third quadrant for photodiode detectors

Negative properties of photovoltaic devices

1. Temperature variation affects the open-circuit output voltage V_{OC} considerably.

Application of photovoltaic devices

1. In watches and calculators.
2. Can provide power in remote industrial facilities.
3. Can be used as power source for space aircrafts.
4. Can be used in energy requirement applications.

4.5   Photodiode Detector

Photodiode detector is a *pn* junction (or PIN) diode used with −ve bias for detection of light intensity, as clear in the third quadrant of I-V. curve (Fig. 4.8). Its property of dependence of reverse bias leakage current on light is used as light detector (Fig. 4.10b). Here, reverse bias leakage minority current (I_S) α light intensity (I_L).

4.5.1   Construction of Photodiode Detector

These are made of semiconductor materials and the effective light entry area of a photodiode is about 0.2 mm^2. Generally, silicon with spectral response peaking near a wavelength (λ) of 0.8 µm is used as light-sensing material, but for optimum response in the infrared region, germanium may be employed.

Schematic diagram of a photodiode is shown in Fig. 4.9a–e. An ordinary *pn* junction can be used as a basic photodiode, but PIN junction provides more satisfactory performance with middle *i* layer being intrinsic, as an intrinsic layer has *V* high generation/recombination rate of *e–h* pairs. The main requirement of a photodiode is to ensure maximum amount of light to reach the intrinsic layer. The substrate (the base semiconductor) is n$^+$ (heavily *n* doped). Light can enter from the top thin p layer (Fig. 4.9a) or from the side in mesa structure (Fig. 4.9b). It is packaged with transparent window (Fig. 4.9c). A schematic circuit with photodiode detector with reverse bias is shown in Fig. 4.9d. The symbol is given in Fig. 4.9e.

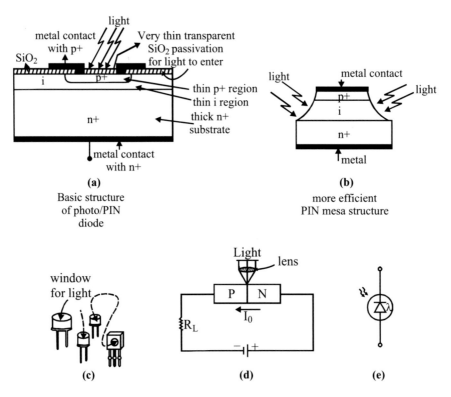

Fig. 4.9 a Basic structure of photo/PIN diode, **b** high-efficient PIN mesa structure for light entering directly into region, **c** packaged photodiodes with windows, **d** schematic circuit of photodiode detectors with reverse biased, **e** symbol of photodiode detector

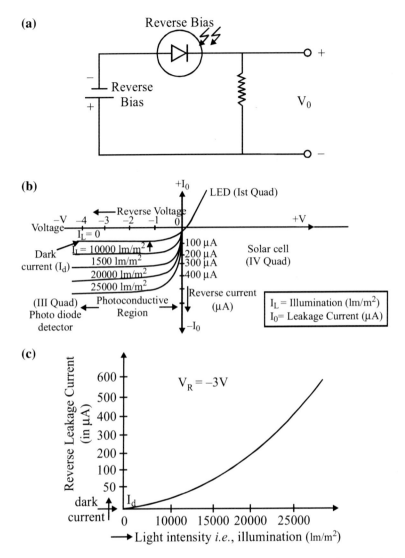

Fig. 4.10 a Circuit connection of photodiode detector for seeing I versus V characteristic. **b** V-I characteristics of photodiode detector where I_L = Illumination (lm/m^2) and I_0 = Leakage current (μA). **c** Illumination versus reverse leakage current characteristics of photodiode detector for $V_R = -3$ V

4.5.2 Working of Photodiode Detector

When a photodiode is reverse biased, there is a leakage current in μA range. This current increases, when light is incident on the i region of the diode due to extra generation of electron–hole pair (EHP) as shown in Fig. 4.10b, c.

Table 4.2 Different materials used in photodiode detectors and their wavelength sensitivity

Material	Wavelength sensitivity (μm)	Colour of light
Germanium (Ge)	0.8–1.7	Infrared (IR)
InGaAs	0.8–1.6	Infrared (IR)
PbS	1.0–3.5	Infrared (IR)
Silicon (Si)	0.19–1.1	Complete optical range + IR + UV

Table 4.3 A typical photodiode detector specifications at 300°k

P_D	V_{DC}	Reverse breakdown voltage (max)	$I_{D\text{-}max}$ (Dark)	Light current at 3.5 lux; $V_R = -2$ V	t_{res}	t_p
100 mW	350 mV	100 V	2 nA	10 μA	2 ns	900 nm (infrared)

Figure 4.10a shows the photodiode connected with reverse bias, whereas Fig. 4.10b shows the I-V characteristics with different intensities of light. The diode reverse bias leakage current versus light intensity characteristics is shown in Fig. 4.10c. Table 4.2 shows the different materials used in photodiodes with its wavelength and colour of light. It can be analysed from Fig. 4.10b, c that for a given reverse bias voltage, the photocurrent (reverse current) I_0 increases with increase in light illumination. The dark current (I_d) is the leakage current when no light is incident. A photodiode can switch its current ON and OFF in nanoseconds (10^{-9} s), and hence its switching speed is highest (Table 4.3).

4.5.3 Properties of Photodiode Detector

Positive properties

1. Linearity is good.
2. Fast switching time.
3. Better frequency response.
4. Excellent spectral response.

Negative properties

1. Small active area.
2. Dark current, *i.e.* leakage current increases with temperature.

4.5.4 Application of Photodiode Detector

1. Detection of both visible and invisible spectrum of lights. (IR + UV + optical radiations).
2. Switching application.
3. Optical communication.
4. Encoders and decoders.
5. Logic circuits with required stability and high speed.

4.6 Phototransistor Detector

A phototransistor detector is a device, when light is incident on the base–emitter junction, it generates electron–hole pair which then increases the current. This current is then amplified by transistor action. Thus, the sensitivity of phototransistors can be increased as much as 10 times more than that of photodiodes detectors, due to its transistor action, with base not connected anywhere (*i.e.* kept hanging).

4.6.1 Construction of Phototransistor Detector

Figure 4.11a, b represents the cross-section and planar look of phototransistor detector.

4.6.2 Working of Phototransistor Detector

The emitter junction (J_E) is kept slightly forward biased and junction (J_C) is reverse biased for transistor action of amplification of base current.

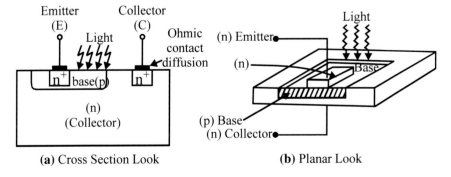

(a) Cross Section Look   **(b)** Planar Look

Fig. 4.11 Structure of phototransistor **a** cross-section look **b** planer look

Many a times, the base connection is kept hanging. In both cases, the illumination on the base region generates extra electron–hole pairs (EHP) which lowers the barrier potential across both emitter–base junction (J_E) and base–collector junction (J_C). Due to it, higher current flow through emitter and collector junction. This emitter current is multiplied with the current amplification factor ($\beta \approx 100$) of the transistor. Therefore, the phototransistor detector provides a much larger output as compared to photodiode detector output. Figure 4.12a–c shows the basic symbol,

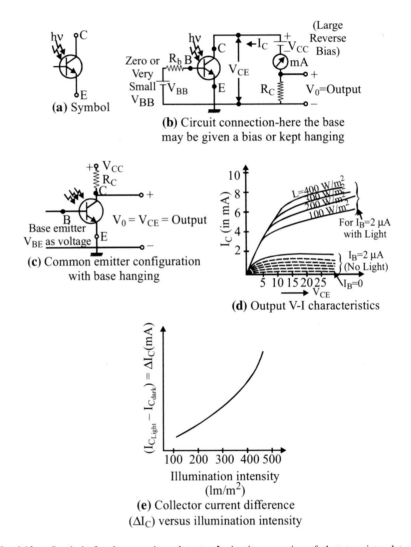

Fig. 4.12 **a** Symbol of a phototransistor detector, **b** circuit connection of phototransistor detector with bias, **c** many a times, the base for CE configuration is kept hanging (not connected) and it still does the same work **d** output V-I characteristics, **e** output current difference (ΔI_C) versus illumination intensity characteristics

circuit configuration with emitter–base junction given bias and circuit configuration with common emitter with base hanging cases, respectively. For no light and with light, the transistor characteristics of I_C versus V_{CE} are shown in Fig. 4.12d. The current illumination characteristics are shown in Fig. 4.12e.

4.6.3 Properties of Phototransistor Detector

Positive properties

1. High gain (β times of photodiode, *i.e.* 100 times).
2. Fast switching time (in pico sec).
3. Small rise and fall time.
4. Extremely stable operation.

Negative property

1. Reverse saturation current, *i.e.* leakage current increases with temperature.

4.6.4 Application of Phototransistor Detector

1. In digital circuits.
2. In optoisolators and photo interrupters.
3. As light-sensitive switches.
4. As a control element to activate various types of load.

Example 4.5 Why a reverse bias GaAs *pn* junction is not a good photodetector of light at $\lambda = 10^{-4} \mu m$.

Solution For GaAs, E_g = 1.43 eV (see Table 4.1 also)
Photon energy must be greater than the energy gap (E_g) for absorption of photon. Here,

$$E_{photon} = \frac{hc}{\lambda} = \frac{4.14 \times 10^{-5} \times 3 \times 10^{10}}{10^{-4}} = 1.24\,eV$$

i.e. $E_{photon} < E_g$
It is not satisfying the condition for absorption of photon, *i.e.* $E_g < E_{photon}$.

Therefore, the GaAs *pn* junction device is not a good photodetector for light since photon will not get absorbed. Therefore, normally $Si(E_g \approx 1.1\,eV)$ or $Ge(E_g \approx 0.7\,eV)$ can be used for $\lambda = 10^{-4} \mu m$ light, where $E_{photon} > E_g$ is satisfied.

4.7   Optocoupler: High Voltage Line Controlled Optically

An optocoupler (optoelectronic coupler) is essentially a combination of photo-transistor detector and LED in one package. Figure 4.13 shows the typical circuit and terminal arrangement for one such device contained in a dual in (DIP) package. When current flows in the LED, the emitted light is directed to the phototransistor detector producing current flow in the transistor. The coupler may be operated as a switch, in which case both the LED and the phototransistor are normally off. A pulse of current through the LED causes the transistor to be switched ON for the duration of the pulse. Linear signal coupling is also possible. Due to the optical coupling, there is a high degree of the electrical isolation between the input and output terminals, and so the term optoisolator is also used. The output (detector) stage has no effect on the input, and the electrical isolation allows a low-voltage dc source to control high-voltage circuits.

The cross-section diagram of Fig. 4.13c illustrates the construction of the op-tocoupler. The LED light as emitter and transistor as detector are connected in a transparent insulating material, which allows the passage of illumination, while maintaining electrical isolation.

Specification

The partial specification for a typical optocoupler is shown in Table 4.4 having three parts a, b, c. The first part specifies the current and voltage condition for the input (LED) stage. The second gives the biasing specifications of the output (phototransistor detector). The third part defines the coupling parameters. The transistor collector current (I_C) is listed as 5 mA (typical) when its $V_{CE} = 10$ V and the LED has $I_E = 10$ mA. In this particular case, the ratio of output current to input current is 50% and this is called as current transfer ratio (CTR) and for an opto-electronic coupler with a transistor output, it can range from 10 to 150%.

Working

The circuit of an optocoupler in a dc- or pulse-type coupling of 24 V system with 4 V system, as shown in Fig. 4.14. When the LED (D_1) diode current is switched ON

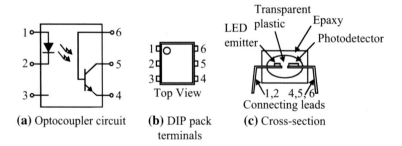

(a) Optocoupler circuit   (b) DIP pack   (c) Cross-section
                             terminals

Fig. 4.13 **a** Optocoupler circuit (composed of LED and phototransistor in a DIP package), **b** DIP packaged optocoupler circuit, **c** input and output of optocoupler are electrically isolated

Table 4.4 Specification of a typical optocoupler **a** input port (LED), **b** output port (phototransistor), **c** coupling parameters

(a) Input port (LED)

I_F (max)	$V_{F(max)}$ (I_F = 20 mA)		V_R
60 mA	1.5 V		3 V

(b) Output port (phototransistor detector)

$V_{CE\ (max)}$	$I_{C(max)}$	P_D	V_{CE} (sat)	I_{CEO} (Dark)
30 V	150 mA	150 mW	0.2 V	50 nA

(c) Coupling parameters

$I_{C(out)}$ [I_F = 10 mA]	$T_{ON(max)}$		$t_{off(max)}$	Isolation voltage
5 mA	2.5 µs		4.0 µs	7500 V

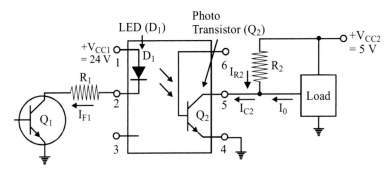

Fig. 4.14 Optoelectric coupler (Optocoupler) used for coupling a signal from a 24 V system to a 5 V systems which are not connected

by the action of the transistor Q_1 operating from a 24 V supply. This LED sends light to transistor Q_2, which gets turned ON into saturation. The collector current of Q_2 provides the load (sinking) current and the current through the pull-up resistor (R_2). This pull-up resistor (R_2) is necessary to ensure that, the load terminal is held at the 5 V supply level, when Q_2 is OFF.

4.8 Light-Dependent Resistor (LDR) or Photoconductive Cell

Light-dependent resistor (LDR) is also called photoconductive cell. It is a device whose resistance decreases with the increase in light intensity. It offers very high resistance at no light condition and very low resistance when light falls on conductive materials. Figure 4.15a, b shows the symbol and top view of an LDR.

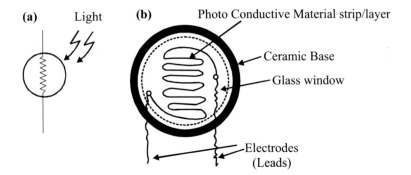

Fig. 4.15 **a** Symbol of LDR, **b** top view of an LDR

4.8.1 Construction and Working Principle of LDR

Light-dependent resistance (LDR) or photoconductive cell is basically made of photoconductive material typically cadmium sulphide, cadmium selenide or cadmium sulpho-selenide, etc, a thin layer of which is deposited on an insulating ceramic base in a zigzag pattern in order to obtain a desired value of resistance and power rating (see Fig. 4.15). The output is taken from the leads coming out of the base. This assembly is enclosed in a metal case, with a glass window over the photoconductive material, having diameter of 0.125″ to 1″.

Figure 4.16a, b gives a simple circuit in which the photoconductive cell (or LDR) is connected in a simple circuit and its illumination characteristics.

(a) **At no light condition**: LDR offers very high resistance when no light is imposed on the window; therefore, the current is very small (see Fig. 4.16, where the log–log scale may be noted).

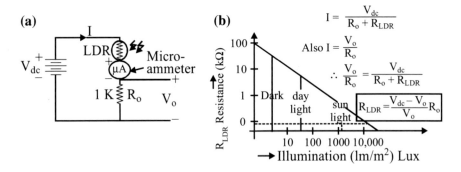

Fig. 4.16 **a** LRD connected in a simple circuit, **b** Illumination characteristics of LDR

(b) **When exposed to light**: In this condition, its resistance decreases which results in increase of current. When light falls on these photoconductive material 'e'–'h' pairs gets generated as the photon energy excites the 'e' to go to conduction band from valance band.

LDR is operated with a series resistance for its protection from reaching heavy current, so that it does not get burnt off. The resistance of LDR is normally around lesser than hundred kilo ohm (100 kΩ) at no light condition in pitch dark and falls to as low as 0.1 Ω with illumination of 10,000 lm/m^2 (*i.e.* sunlight of Asia).

4.8.2 Properties of Light-Dependent Resistor (LDR)

Positive properties

1. Offer very high resistance with no light (*i.e.* also called as dark resistance) of the order of 10^6 Ω in pitch dark room to 10 Ω in bright sunlight. CdS responds to visible light, while CdSe responds from yellow to IR.
2. Available in different sizes.
3. Light in weight.
4. Low power dissipation.

Negative properties

1. Sensitive to variations of temperature. CdS is most stable with temperature.
2. Operate only at limited current ranges (<10 mA to 10 μA).
3. Time of response is large, e.g. 100 m sec in CdS to 10 m sec in CdSe.

4.8.3 Applications of Light-Dependent Resistor (LDR)

1. In toys.
2. In smoke detector circuits.
3. In clocks.
4. For automatic brightness compensation in TV receivers and desktop computer.
5. Light control switch.
6. In remote control switch of TV.
7. For volume control in audio and power amplifiers.

Example 4.6 In given Fig. 4.17, an LDR has resistance of only 15 Ω at a certain very high light intensity. What value of a series resistance required if a current of no more than 10 mA is to flow with a supply voltage of 9 V?

Fig. 4.17 LDR Example 4.6

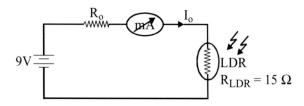

Solution Maximum allowed current through $LDR = 0.01$ A

$$\therefore \quad V_{dc} = I_0(R_0 + R_{LDR})$$

$$\therefore \quad \frac{9}{0.01} = (R_0 + 15)$$

$$900 - 15 = R_0$$

$$R_0 = 885\,\Omega$$

Here protection resistance is 885 Ω **Ans.**

4.9 Infrared Emitters

The infrared emitters are GaAs–LED and are *pn* junction devices, which emit the beam (packets) of light, when forward biased (see Table 4.1 and Fig. 4.1). It works on the principle of LED (see art 4.1 and 4.2). When a GaAs–*pn* junction is forward biased then some of the electrons of the valence band (E_V) absorbs energy/get excited and jumps the band gap of $E_g = 1.4$ eV to reach the conduction band (E_C), creating a hole in the valence band. At the same time, some of the excited electron keeps returning back to the valence band (E_V) to recombine with the holes they have left them. In this process, these returning electrons loose energy (*i.e.* emits earlier absorbed energy), frequency of which is given by $E_g = hv$. For GaAs, the frequency (v) is the range of infrared frequency (also seen in Figs. 4.1, 4.2 and Table 4.1).

Thus, we see that we get a LED in infrared region by using GaAs *pn* junction, which is called infrared emitters. The radiant energy from the device is infrared with a typical peak at $\lambda = 0.9$ μm. The optocouplers also use these infrared LED emitters of $\lambda = 1.0$ μm, using silicon photodiode and silicon phototransistors as receivers (see Fig. 4.1).

These infrared emitting diodes are widely used in shaft encoder, data transmission systems, intrusion alarms, card and paper tape readers, and high-density mounting application. The shaft encoder can produce 150 μW of radiant energy at 1.2 V and 50 mA.

4.10   Liquid Crystal Display (LCD)

It is one of the most fascinating materials having the properties of liquids as well as solid crystal. They have crystalline arrangement of molecules, yet they flow like a liquid. LCD does not emit or generate light but alters the light falling on it. The property of modulating the light falling on it is only when electric field is applied across the liquid by putting a voltage across. This is because of disruption of its originally well-ordered structure. The LCD molecules are rod-like in shape of length   20–130   A°U.   Basically,   LCD   is   an   organic   material   of   N-4 Methoxy-Benzy-lidere-4 Butaline (Fig. 4.18), and they come back to the original alignment after removal of field.

These rod-like molecules (called directors) in the original crystal have different arrangements in the following three types of liquid crystals (LC):

1. **Smectic**: Figure 4.19a shows smectic structure of liquid crystals. In this struc-ture, the rod-like molecules are arranged in layers, and within each layer there is a fixed vertical orientational order. Thus, in a given layer, the rods are all oriented in the same direction.
2. **Nematic**: Figure 4.19b shows nematic structure of liquid crystals. In the nematic structure, the positional order between layers of molecules is lost, but the vertical orientation order is maintained.
3. **Cholesteric**: Figure 4.19c shows cholesteric structure of liquid crystals. In these crystals, the rod-like molecules in each layer are oriented at different angle.

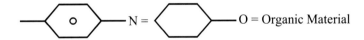

Fig. 4.18   Structure of LCD molecule

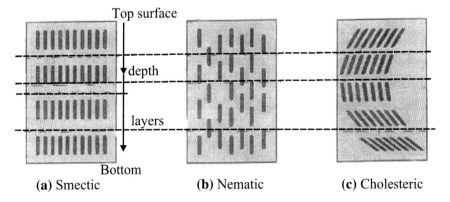

(a) Smectic   **(b)** Nematic   **(c)** Cholesteric

Fig. 4.19   The three different structures of the liquid crystal (LC) molecules (directors)

Within each layer, the orientation order is maintained. The cholesteric liquid crystal is related to the nematic crystal with the difference being the twist of the molecules as one goes from one layer to another.

The optical activity of the crystal depends on the orientation and the twist of the molecules as one goes from one layer to another.

Types of LCDs

There are two types of liquid crystal displays (LCDs) according to the theory of operation, where any of the above three types of liquid crystal can be used:

1. Dynamic scattering
2. Field effect.

    1. **Dynamic scattering type LCDs**: It gives light display on dark background. Figure 4.20 shows the construction of the liquid crystal display (LCD), consisting of two glass plates with liquid crystal (LC) fluid in between which is 5–50 μm thick. The bottom glass plate is coated (at the liquid crystal side) with the transparent layer of the conducting material (SnO_2), which makes the bottom reflecting mirror for light also. The front glass is also fully coated at the liquid crystal side with conducting transparent layer of SnO_2, which is etched so as to remain the seven-segment pattern along with its leads and seven-segment contact pads of each digit. The remaining SiO_2 is etched out (see Fig. 4.21).

Operation: Figure 4.21 shows the operation of an LCD. In the absence of the electric field, original orientation order (Fig. 4.19) is maintained, which allows the outside light to enter. As the electric field is applied on the elements, the current through the LC causes disturbance of the vertical orientation order of directors, resulting in scattering of light entering it and that element becomes visible in the dark background (where no field is applied) (Fig. 4.22).

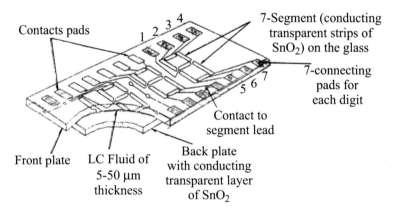

Fig. 4.20 Liquid crystal display construction

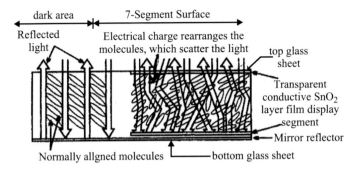

Fig. 4.21 Dynamic scattering mechanism in the liquid crystal (LC)

Fig. 4.22 Typical liquid crystal of dynamic scattering LCD

2. **Field-effect LC display**: It gives dark display on light background. In these displays, nematic normal liquid crystals are used. Figure 4.23 shows operation of field-effect liquid crystal display with nematic crystals. It consists of two glass plates, a liquid crystal fluid polarizers and transparent conductors. The liquid crystal fluid is sandwiched between two glass plates. Each glass plate is associated with light polarizer. The light polarizers are placed at right angle to each other. In the absence of electrical excitation of the seven-segment elements, the light coming through the polarizer is rotated 90° in the fluid (LC) and passed through the rear polarizer. It is then reflected back to the viewer by the back mirror (as shown in Fig. 4.23a) after getting re-rotated by −90°.

On the application of electrostatic field, the liquid crystal fluid molecules get aligned, and therefore light through the molecules is not rotated by 90° and it is blocked by front polarizer as shown in Fig. 4.23b. This causes the appearance of dark digit on a background as shown in Fig. 4.23c.

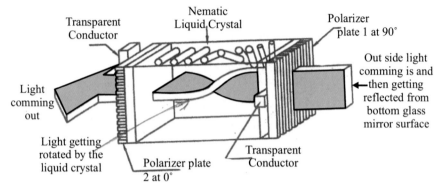

(a) Field Effect Display "ON State" in absence of electric field

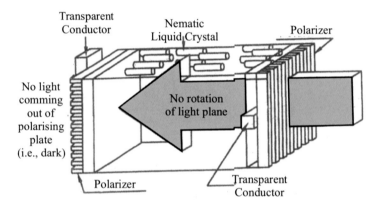

(b) Field Effect Display "OFF State" in electric field

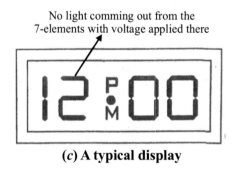

(c) A typical display

Fig. 4.23 Field-effect LC display **a** without field **b** with field **c** typical liquid crystal display

4.10.1   Properties of Liquid Crystal Display (LCD)

Positive properties

1. Voltage required is very small.
2. Low power consumption.
3. Economical.

Negative properties

1. Very slow devices.
2. Turn ON and OFF durations are quite large.
3. When used on d.c., their life span is very small.
4. They occupy large area.

4.11   Comparison Between LED and LCD

Table 4.5 gives the comparison between LED and LCD.

Summary Sheet (Photoelectric Devices)

Using semiconductor and photoconductive materials

These devices can be classified into the following three groups:

1. **LED**: Some semiconductor diodes in forward bias emit light of certain colours as the band gap energy is in the optical region as

$$E_g = hv$$

(These are GaAs, GaP, GaAsP, etc.)
2. **Solar cells or photovoltaic cell (no bias required)**: When light falls on a *pn* junction then voltage is generated across the junction due to accumulation of EHP (excess holes on *p* side and excess electrons on *n* side).
3. **Photodetectors of light**: Detection of light is done by using (a) photoconductive material (e.g. CdS, CdSe, Cd–Se–S), (b) Si diode or (c) PIN diode and (d) phototransistor.
   (a) **Light-dependent resistors (LDR)-photoconductive cell**: Here the photoconductive material is used and intensity of light can be measured by the resistance, which changes with light.
   (b) **Photodiode detectors (*pn* type)**: In reverse-biased diode, light falling on the junction increases the leakage current to a large extent by generating EHP. The change in the leakage current indicates the intensity of light.

Table 4.5 Gives the comparison between LED and LCD

S. no.	Parameter	LED	Dynamic scattering LCD	Field-effect LCD
1.	Power/digit (P_a)	Very high 10–140 mW and depending on colour	100 µW	1–10 µW
2.	Voltage	5 V	18 V	3–7 V
3.	Temperature range (°C)	−55 to 125	0–80	0–70
4.	Switching speed	1 µs (Very fast)	300 µs	100–300 µs
5.	Life in hours	100,000+	10,000	10,000
6.	Colours of letters numerals	Red, orange, yellow, green	white in black background	black in white background
7.	Brightness	Good to excellent	Average	Average
8.	Contrast ratio	10:1	20:1	20:1
9.	Appearance	Good to excellent	Good	Good to excellent
10.	Viewing angle	150°	90–150°	90–120°
11.	Font	7 and 16 segment 5 × 7 dot matrix	7 and 16 segment	7 and 16 segment
12.	Vertical size	0.1–0.6 in.	0.2–8 in.	0.2–2 in.
13.	MOS compatibility	Small yes large no, as Power/m² is high. It requires external circuit for power	Yes, it can be driven from IC-chip supply itself	Yes, It also can be driven from IC-chip power supply
14.	Cost per digit (0.3–0.6 10.)	25 Rs.	25 Rs.	25 Rs.
15.	Ruggedness	Excellent	Good	Good
16.	Ease of mounting	Excellent	Poor	Poor
17.	Applications	1. Seven-segment display 2. Indicator 3. Matrix displays	1. Seven-segment display 2. Calculators, pagers and wristwatch 3. Flat-panel displays 4. Alphanumeric display	

(c) **Phototransistor detector**: In Si transistor (BJT) if the base region is exposed/open to light, then the

  (i) leakage current of the reverse-biased collector junction increases.
 (ii) minority current of emitter junction increases as more number of e–h pairs get generated. This extra e–h pair (EHP) gets multiplied by β of the transistor gain, due to transistor action. This leads to larger collector current.

4. **Optocouplers**: The phototransistor's current is controlled by IR or optical light by light of an LED. Thus, a small voltage at LED can trigger very large voltage at the output of this single device packed in one box.
5. **The three types of functions of diodes are as follows**:
   1. LED.
   2. Solar cell or photoelectric cell.
   3. Photodiode detector.

Questions

Fill in the blanks

1. In photoconductive cells, _____ alloys are used.

2. The photovoltaic cell consists of _____ layers.

3. Usually LDR is made up of single _____.

4. A solar cell is an example of _____.

5. In photovoltaic cell, light is directly converted to _____.

6. LED works on _____.

7. Operating range of LED is _____volts bias.

8. The LED is a *pn* junction diode which emits light when _____.

9. In a photodiode, when no light is applied, the minimum leakage current is called as _____.

10. In photodiodes, photon's energy is absorbed by the _____.

11. A solar cell works _____.

12. The resistance of photoresistor in dark can go up to _____.

13. The viewing angle of LCD is _____ and for LED, it is _____.

14. LED requires _____ power per digit than LCD.

15. The isolation resistance of an optoisolator is in the order of _____.

Short Questions

1. What is optoelectric devices?

2. With the help of neat sketches, explain about photoconductive cell.

3. With help of neat sketches, explain about photovoltaic cell.

4. Explain about liquid crystal display (LCD).

5. With the help of neat sketches, explain about photodiode.

6. With the help of neat sketches, explain about phototransistors.

7. Write short notes on LED.

8. Explain the working of infrared emitters.

9. Discuss the working the optocouplers.

10. Write the differences between LED and LCD.

11. List all the advantages and disadvantages of LCD.

12. What are the applications of photodiode and phototransistors.

13. Write the differences between photoconductive cell and photovoltaic cell.

14. List out the applications of solar cell.

15. Briefly explain the principle and working of a solar cell.

16. Write short notes on
    (i) LDR.

    (ii) Optocouplers.

Long Questions

1. With help of neat sketches, explain the working of photoconductive cell, photovoltaic cell and solar cell.

2. With help of neat sketches, explain the working of photodiode and photo-transistors. Explain their working principles and their differences.

3. Discuss about the optoisolators, whatever you know.

4. Highlight the characteristics of photodiode and solar cell.

5. Explain the LED, infrared emitters and photodiode, structure and working.

6. Along with a diagram, explain the working of a phototransistor.

7. Describe the principle involved in phototransistor and light-dependent resistor.

8. Describe in detail the principle of operation of LED and LCD, with neat diagrams. Also compare LED and LCD.

9. Elaborate on different types of display devices giving advantages and disadvantages of each.

10. Discuss the dynamic scattering type of LCD.

11. Discuss the field-effect type of LCD. What other classifications are there for LCD? Discuss.

15. 'Si diode can function as LED in infrared region'. Explain

16. 'Life longevity of LED is very very long as compared to any other diodes'. Why?

17. What are the inputs given to a photoconductive cell/solar cell and what are the outputs?.

Chapter 5
Digital Electronics

Contents

© Springer Nature Singapore Pte Ltd. 2020
S. S. Srikant and P. K. Chaturvedi, *Basic Electronics Engineering*,
https://doi.org/10.1007/978-981-13-7414-2_5

5.1 Introduction

A digital circuit/signal is different from analog circuit/signal. The term digital is derived from function of circuits which perform operations of counting digits, etc. Applications of digital electronics are in computer system, telephony, data processing, communication systems, radar navigation, military systems, medical instruments, consumer products, toys, etc. In the modern time, the digital technology plays a very important role in the daily life and has been progressed from IC to VLSI/ULSI on chips. Digital circuits involve operations in which there are only two possible states.

5.2 Number System

The system used to count discrete units (or articles) is called number systems. The decimal number system has cycle of 10 digits *i.e.* 0 to 9, while the digital number system has cycle of 2 digits only, *i.e.* 0 and 1 called binary. There are three number systems which are often used in digital electronics: (*a*) binary number system, (*b*) octal number system and (*c*) hexadecimal number system.

The binary number cycles after two digits 0, 1, 10, 11, 100, etc. while octal cycles back after every 8 digits, e.g. 0, 1, 2, 3, …, 7, 10, 11, 12, …, 17, 20, 21, …, etc.

5.2.1 Binary Numbers

Binary number system uses only two digits, *i.e.* 0 and 1. For any number system, it is represented mathematically as

(Number)$_{Base}$

For decimal number system, Number → 0 to 9, *i.e.* (Number)$_{10}$ and Base → 10.

For binary number system, Number → 0 to 1, *i.e.* (Number)$_2$ and Base → 2.

The position of 1 or 0 in a binary number indicates its 'weight' or place value within the number. The weight of each successively higher position (to the left) in a binary number is an increasing power of two. For example, in decimal system

$$(138)_{10} = \underset{\text{Hundreds}}{1 \times 10^2} + \underset{\text{Tens}}{3 \times 10^1} + \underset{\text{Ones}}{8 \times 10^0}$$

Similarly, the binary numbers are also represented by positional weights. For example,

$$(138)_{10} = (10001010)_2$$
$$= 1 \times 2^7 + 0 \times 2^6 + 0 \times 2^5 + 0 \times 2^4 + 1 \times 2^3 + 0 \times 2^2 + 1 \times 2^1 + 0 \times 2^0$$
$$= 128 + 0 + 0 + 0 + 8 + 0 + 2 + 0$$
$$= 138$$

In the digital system, each digit, either 0 or 1, is referred to as a bit. A string of four bits is called a nibble and eight bits make byte. Hence, 1101 is a binary nibble and 10110101 is a binary byte of two nibbles.

Note: The symbol to separate the integer part from fractional part of a number is called **radix point**. This radix is called decimal point for 0–9 number system.

5.2.2 Decimal to Binary Conversion

To convert decimal number into binary number, the integer part (*IP*) and fractional part (*FP*) of the number should be considered separately.

Double-dibble method: This method is the most commonly used for converting decimal number into their binary equivalents. The integer part (*IP*) of the decimal system is progressively divided by 2 and the remainders after each division noted till the quotient becomes '0'. The remainders written in the reverse order, constitute the binary equivalent.

For the conversion of the fractional part (FP), the fraction is progressively multiplied by 2 and the carry of digit to the left of decimal is recorded, till the result of the multiplication becomes '0' with a carry of '1' or the desired number of digits have been obtained whichever is earlier. The carries written in the forward order then constitute the binary equivalent of fractional part.

Example 5.1 Find the binary equivalent of $(16.25)_{10}$.

Solution For integral part (IP), divide by 2

Divisor	Integer Part	Remainder
2	16	—×—
2	8	0 (LSB)
2	4	0
2	2	0
2	1	0
2	0	1 (MSB)

Binary Equivalent

The remainders written in the reverse order will be binary equivalent of integer part (IP) = 10000.

For fractional part (FP), multiply by 2

$0.25 \times 2 = 0.5$ with a carry of full digit $= 0$ $\left.\right|$ Fractional
$0.5 \times 2 = 1.0$ with a carry of full digit $= 1$ $\left.\right\downarrow$ Equivalent

The carry written in the forward order will be binary equivalent of fractional part (FP).

Fractions binary equivalent $= 0.01$

∴ Binary equivalent of (16.25) is therefore (IP). (FP), *i.e.* 10000.01

∴ $(16.25)_{10} = (10000.01)_2$

Example 5.2 Convert decimal number 13.73 into binary equivalent up to 5-decimal point.

Solution For integer part, we divide by 2,

Divisor	Integer Part	Remainder
2	13	$-\times-$
2	6	1 (LSB)
2	3	0
2	1	1
2	0	1 (MSB)

LSB

↑

Binary Equivalent

MSB

Binary equivalent of 13 = 1101.
For fractional part, we multiply by 2

$0.73 \times 2 = 0.46$ with carry 1 | MSB
$0.46 \times 2 = 0.92$ with carry 0
$0.92 \times 2 = 0.84$ with carry 1   Binary Equivalent
$0.84 \times 2 = 0.68$ with carry 1
$0.68 \times 2 = 0.36$ with carry 1 ↓LSB (upto 5-decimal digit)

Binary equivalent $0.73 = 0.10111$

$$(13.73)_{10} = [(IP) \cdot (FP)]_2 = (1101 \cdot 10111)_2$$

Example 5.3 Convert $(320)_{10}$ into binary number system.

Solution $(320)_{10} = ( )_2$

2	320	—	$\times$ —
2	160	—	0 (LSB)
2	80	—	0
2	40	—	0
2	20	—	0
2	10	—	0
2	5	—	0
2	2	—	1
2	1	—	0
	0	—	1 (MSB)

(reading from bottom to top)

MSB

$$(320)_{10} = (101000000)_2$$

Example 5.4 Convert the decimal number $(250.5)_{10}$ into base 3, base 7 numbers. Check the answer by reverse process, *i.e.* back to decimal.

Solution For Base 3: $(250.5)_{10} = (100021.1111)_3$ (**Please verify for practice**)

$$(100021)_3 = 1 \times 3^5 + 2 \times 3^1 + 1 \times 3^0 = 343 + 2 \times 3^1 + 1 \times 3^0 = 243 + 6 + 1 = (250)_{10}$$
$$(0.1111)_3 = 1 \times 3^{-1} + 1 \times 3^{-2} + 1 \times 3^{-3} + 1 \times 3^{-4}\dots \cong (0.5)_{10}$$

For Base 7; $(250.5)_{10} = (505.333)_7$ (**Please verify for practice**)

$$(505)_7 = 5 \times 7^2 + 0 \times 7^1 + 5 \times 7^0 = 250$$
$$(0.333)_7 = 3 \times 7^{-1} + 3 \times 7^{-2} + 3 \times 7^{-3} = 0.4999 \approx 0.5$$

5.2.3 Octal Numbers

The octal number systems have a radix or base of eight which means that it has eight distinct digits. These different digits are 0, 1, 2, 3, 4, 5, 6 and 7. All higher digits are expressed as a combination of these on the same pattern as the one followed in case of binary and decimal number system.

Example 5.5 Convert $(12)_{10}$ into octal number

8	12		
8	1	—	4
	0	—	1

Octal equivalent = 14

$$(12)_{10} = (14)_8$$

Example 5.6 Convert $(570)_{10}$ into octal number

Divisor	I.P.	Remainder
8	570	—
8	71	2
8	8	7
8	1	0
	0	1

LSB

Octal equivalent = 1072

MSB

$$\therefore \qquad (570)_{10} = (1072)_8$$

Example 5.7 Convert $(1052)_8$ to decimal

Solution

$$(1052)_8 = 1 \times 8^3 + 0 \times 8^2 + 5 \times 8^1 + 2 \times 8^0$$
$$= 512 + 0 + 40 + 2$$
$$= (554)_{10} \quad \textbf{Ans.}$$

5.2.4 Hexadecimal Numbers

Hexadecimal number system has a radix of 16 and uses 16 symbols, namely, 1, 2, 3, 4, 5, 6, 7, 8, 9, A, B, C, D, E and F (see Table 5.1). The symbols A, B, C, D, E and F represent the decimals numbers 10, 11, 12, 13, 14 and 15, respectively, *i.e.* the number system is recycled after every 16 digits, *i.e.* after F, we have $(10)_{16}$. The place values or weights of different digits in a hexadecimal are as given below:

$...x_1$	x_2	x_3	x_4	$\bullet$	x_5	x_6	x_7	$x_8...$
16^3	16^2	16^1	16^0		16^{-1}	16^{-2}	16^{-3}	16^{-4}

Radix point

Here $x_1, x_2, \ldots$ are hexadecimal digits.

Example 5.8 Convert $(235.25)_{10}$ into hexadecimal.

Solution Integral part

16	235	Remainder	
	14	11 (B in Hexadecimal)	Hexadecimal
	0	14 (E in Hexadecimal)	equivalent

$$\therefore \quad (235)_{10} = (EB)_{16}$$

Fractional part

$$0.25 \times 16 = 0 \quad \text{with a carry } 4 \downarrow$$
$$\therefore \qquad (0.25)_{10} = (0.4)_{16}$$
$$\therefore \quad (235.25)_{10} = (EB.4)_{16}$$

Table 5.1 Equivalence of decimal with binary, octal, hexadecimal and BCD systems

S.N.	Decimal	Binary	Octal	Hexa-decimal	Binary Coded Decimal (BCD)
1	0	00	0	0	0000
2	1	01	1	1	0001
3	2	$(10)_2 = (2)_{10}$	2	2	0010
4	3	11	3	3	0011
5	4	$(100)_2 = (4)_{10}$	4	4	0100
6	5	101	5	5	0101
7	6	110	6	6	0110
8	7	111	7	7	0111
9	8	$(1000)_2 = (8)_{10}$	10	8	1000
10	9	1001	11	9	1001
11	10	1010	12	A	$(0001\ 0000)_{BCD} = (10)_{10}$
12	11	1011	13	B	0001 0001
13	12	1100	14	C	0001 0010
14	13	1101	15	D	0001 0011
15	14	1110	16	E	0001 0100
16	15	1111	17	F	0001 0101
17	16	$(10000)_2 = (16)_{10}$	20	$(10)_{16} = (16)_{10}$	0001 0110
18	17	10001	21	11	0001 0111
19	18	10010	22	12	0001 1000
20	19	10011	23	13	0001 1001
21	20	10100	24	14	$(0010\ 0000)_{BCD} = (20)_{10}$
22	21	10101	25	15	0010 0001
23	22	10110	26	16	0010 0010
24	23	10111	27	17	0010 0011
25	24	11000	30	18	0010 0100
26	25	11001	31	19	0010 0101
27	26	11010	32	1A	0010 0110
28	27	11011	33	1B	0010 0111
29	28	11100	"	1C	0010 1000
30	29	11101	"	1D	0010 1001
31	30	11110	"	1E	$(0011\ 0000)_{BCD} = (30)_{10}$
32	31	11111	"	1F	"
"	32	$(100000)_2 = (32)_{10}$	"	$(20)_{16} = (32)_{10}$	"
"			"	"	"
			"	"	$(1001\ 1001)_{BCD} = (99)_{10}$
				"	$(0001\ 0000\ 0000)_{BCD} = (100)_{10}$
				2F	$(0001\ 0000\ 0001)_{BCD} = (101)_{10}$
				$(30)_{16} = (48)_{10}$	

NB: The line between numbers shows the starting of the next cycle of the system. Decimal after 10th no., Binary after 2nd, 4th, 8th, 16th etc, Octal after every 8th, Hexa after evert 16th, BCD after 10th, 100th, ...

Example 5.9 Convert $(235)_{10}$ into hexadecimal.

Solution

16	256	Remainder
	15	13 (D in Hexadecimal)
	0	15 (F in Hexadecimal)

$$\therefore \quad (253)_{10} = (FD)_{16}$$

Example 5.10 Convert $(EB4A)_{16}$ into decimal.

Solution

A   B   C   D   E   F

(10) (11) (12) (13) (14) (15)

$$(EB4A)_{16} = E \times 16^3 + B \times 16^2 + 4 \times 16^1 + A \times 16^0$$
$$= 14 \times 16^3 + 11 \times 16^2 + 64 + 10 \times 1$$
$$= 57{,}344 + 2816 + 64 + 10$$
$$= (60234)_{10}$$

Example 5.11 Convert $(A13B)_{16}$ into decimal.

Solution

$$\overset{3\ 2\ 1\ 0}{(A\ 1\ 3\ B)_{16}} = A \times 16^3 + 1 \times 16^2 + 3 \times 16^1 + B \times 16^0$$
$$= 10 \times 16^3 + 256 + 48 + 11 \times 1$$
$$= 40960 + 256 + 48 + 11$$
$$= (41275)_{10}$$

5.2.4.1   Hexadecimal—Binary Conversion

Conversion from hexadecimal to binary or vice versa can be easily carried out. For obtaining the binary equivalent of a hexadecimal number, replace each significant digit in the given number by its 4-bit binary equivalent. For example, by Table 5.1:

$$(26 \cdot A)_{16} = \underset{0010}{\overset{2}{}} \quad \underset{1110}{\overset{6}{}} \cdot \underset{1010}{\overset{A}{}}$$

$$\therefore \quad (26 \cdot A)_{16} = (0010\ 0110 \cdot 1010)_2$$

The reverse procedure is used for converting binary to hexadecimal, *i.e.* starting from the LSB, replace each group of 4 bits by their decimal equivalents. Put zeros to left of MSB and zeros to the right of LSB of decimal point if required, as these zeros do not change the value of a number. For example,

$$1110\ 0110\ 1111 = \underset{E}{\underline{1110}}\ \underset{6}{\underline{0110}}\ \underset{F}{\underline{1111}}$$

$$\therefore \qquad (1110\ 0110\ 1111)_2 = (E6F)_{16}\ \text{or}\ (E6F)_H$$

5.2.4.2 Hexadecimal—Octal Conversion

Conversion between hexadecimal and octal is sometimes very much required. Conversion of a hexadecimal to octal, or vice versa is done via decimal number

---

  (i)   Convert the given hexadecimal number to equivalent binary.
 (ii)   Starting from the LSB form groups of 3-bits.
(iii)   Write the equivalent octal number for each group of 3 bits.
 (iv)   Adding zero in LHS of MSB may be required for making 3-bit sets and zero in RHS of LSB of decimal point if required. These zeros do not change the value.

For examples: (*a*) $(35)_{16} = \underset{6}{\underline{11\ 0}}\ \underset{5}{\underline{101}}$

$$= (65)_8$$

(*b* $(8 \cdot 2)_{16} = \underset{2}{\underline{010}}\ \underset{0}{\underline{000}} \cdot \underset{2}{\underline{010}} = (20 \cdot 2)_8$

(*c*) $(2AF)_{16} = 0010\ 1010\ 1111$

$$= \underset{1}{\underline{001}}\ \underset{2}{\underline{010}}\ \underset{5}{\underline{101}}\ \underset{7}{\underline{111}}$$

$$\therefore \qquad (2AF)_{16} = (1257)_8$$

(*d*) $(4A)_{16} = 0100\ 1010$

$$= \underset{1}{\underline{001}}\ \underset{1}{\underline{001}}\ \underset{2}{\underline{010}}$$

$$\therefore \qquad (4A)_{16} = (112)_8$$

To convert an octal number to hexadecimal, the following steps to be followed:

(i)	Convert the given octal number to equivalent binary.
(ii)	Starting from the LSB form groups of 4 bits and add zero in LHS of MSB and zero at RHS of LSB of decimal point if required.
(iii)	What is the equivalent hexadecimal number for each group of 4 bits?

For example $\qquad (37)_8 = 011\quad 111$

$$= \underbrace{0001}_{1}\ \underbrace{1111}_{F} = (1F)_{16}\ \text{(Two zeros added in MSB)}$$

$$(24)_8 = 010\quad 100 = 0001\ 0100\ \text{(Two zeros added in LHS i.e. MSB)}$$
$$= (14)_{16}$$

Thus the rule is
Hexa $\rightleftarrows$ (4-bit binary group) $\rightleftarrows$ (3-bit binary group) $\rightleftarrows$ octal.

5.3   Binary Arithmetic

The arithmetic rules for addition, subtraction, multiplication and division of binary numbers are as follows:

	Addition	Subtraction	Multiplication	Division
(a)	$0 + 0 = 0$	$0 - 0 = 0$	$0 \times 0 = 0$	$0 \div 1 = 0$
(b)	$0 + 1 = 1$	$1 - 0 = 1$	$0 \times 1 = 0$	$1 \div 1 = 1$
(c)	$1 + 0 = 1$	$1 - 1 = 0$	$1 \times 0 = 0$	$1 \div 0 = $ (Not Allowed)
(d)	$1 + 1 = 10$	$10 - 1 = 1$	$1 \times 1 = 1$	$0 \div 0 = $ (Not Allowed)

5.3.1   Binary Addition

The two binary numbers can be added in the same way as two decimal numbers are added. The addition is carried out from LSB (*i.e.* RHS) and proceeded to higher significant bits, adding the carry resulting from previous addition each time. Let us consider an example of the binary number 1011 and 1111.

Carry Carry Carry Carry

(1)   (1)   (1)   (1)   LSB   In   Decimal

    1   0   1   1 $\longrightarrow$ 1   1

+   1   1   1   1 $\longrightarrow$ 1   5

1   1   0   1   0 $\longrightarrow$ 2   6

5.3.2   Binary Subtraction

The binary subtraction is carried out in the same way as that of decimal number. The subtraction is carried out from LSB and proceeded to higher significant bit. The following example explains the steps involved as just like decimal number. Let us consider 1001 is subtracted from 1110.

MSB   LSB

                              (0)   (10) ← Borrow

    1   1   ƚ   0

− 1   0   0   1

    0   1   0   1

Let us consider another example $(1110)_2 - (0111)_2$.

MSB   LSB

    0   (10)   (10)   (10) ← Borrow

    ƚ   ƚ   ƚ   0

− 0   1   1   1

    0   1   1   1

5.3.3   Binary Multiplication

Binary multiplication is rather easier than decimal multiplication. The procedure is same as that of decimal. Let us consider an example 1101 × 0111.

		MSB				LSB
		1	1	0	1
		×	1	1	1
		1	1	0	1
Add this	1	1	0	1	×
Carry	1	1	0	1	×	×
+ (1) (1) (1) (1)	×	×	×
1	0	1	1	0	1	1

5.3.4 *Binary Division*

Division in binary follows the same procedure as division in decimal. Division by O is meaningless and hence not possible.

Example 5.12 (*a*) Divide 110 by 10 (*b*) 1111 by 110.

Solution

$$
\begin{array}{r}
11 \\
(a)\ 10\overline{)110} \\
-10 \\
\hline
10 \\
-10 \\
\hline
00
\end{array}
$$

$$
\begin{array}{r}
10.1 \\
(b)\ 110\overline{)1111.0} \\
-110 \\
\hline
00110 \\
-110 \\
\hline
000
\end{array}
$$

5.4 1's, 2's, 9's and 10's Complements

The usefulness of the complement numbers stems from the fact that subtraction of a number (A) from another (B) can be accomplished by adding the complement of the subtrahend (B) to the minuend (A), followed by minor manipulations of the carry if it comes.

5.4.1 1's Complement Subtraction (M − N)

1's complement of number N to be subtracted is obtained firstly by changing all 1's to 0's and vice versa as shown below:

Binary Number (N)   1's complement of $N = \overline{N}$
1 0 0 1 1 0   0 1 1 0 0 1

Case: (a) M − N when M > N

Step 1. Find the 1's complement form of the N to be subtracted (smaller).
Step 2. Add this 1's complement to M.
Step 3. Remove the carry and add it to the result.

Example 5.13 Subtract $(1001)_2$ from $(1010)_2$ using 1's complement method. Show direct subtraction for comparison.

Solution 1's complement method

$$M - N = (1010)_2 - (1001)_2$$

Here $M > N$

1's complement method

Step 1. 1's complement of $N = \overline{N} = (0110)$

Step 2. $M \rightarrow 1\ 0\ 1\ 0$
          $\overline{N} \rightarrow +0\ 1\ 1\ 0$ **Direct Method**
          Carry $(1)\ \overline{0\ 0\ 0\ 0}$

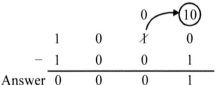

Step 3. Add carry $\rightarrow 0\ 0\ 0\ 0$

1	0	̸1	0
− 1	0	0	1
Answer 0	0	0	1

              $+\quad 1$
          Final answer $\overline{0\ 0\ 0\ 1}$

Case: (b) M − N when M < N

Step 1. Find the 1's complement form of
the N (Larger).
Step 2. Add this to M.
Step 3. Answer is 1's complement of the result with negative sign.

Example 5.14 Subtract $(1110)_2$ from $(1000)_2$ using 1's complement method. Show direct subtraction for comparison.

Solution 1's complement method

$$M - N = (1000)_2 - (1110)_2$$

Here $M < N$

Direct Method

```
  1 0 0 0
  1 1 1 0
 ─────────
 -0 1 1 0
```

Step 1. 1's complement of $N = \overline{N} = 0001$.

Step 2. $M + \overline{N}$

$$
\begin{array}{rl}
M \rightarrow & 1\,0\,0\,0 \\
N \rightarrow & 0\,0\,0\,1 \\
\hline
(0)\, & \overline{1\,0\,0\,1} \\
\end{array}
$$

$\downarrow$

As no carry generated, therefore the result is $-$ve.

Step 3. Get 1's complement of above and put $-$ve sign.

$$\therefore \quad M - N = -(0\,1\,1\,0)_2 \quad \textbf{Ans.}$$

5.4.2   2's Complement Subtraction $(M - N)$

2's complement of a binary is obtained by adding 1 to 1's complements of that number, *i.e.* 2's complement = 1's complement + 1.

Case: (*a*) $M - N$ where $M > N$

Step 1. Find the 2's complement of N (Smaller) $= (\overline{N} + 1) = \overline{N}'$.
Step 2. Add this to M $(M + \overline{N}')$.
Step 3. Discard carry as it will be there, and therefore result will be +ve.

Example 5.15 Subtract $(1010)_2$ from $(1111)_2$ using 2's complement.

Solution $M - N = (1111)_2 - (1010)_2$

Step 1. 2's complement of

$$\overline{N} = (0101)$$
$$+ 1$$
$$\overline{N'} = \overline{N} + 1 = \overline{(0110)}$$

Step 2. Add this to M

$$M = 1\ 1\ 1\ 1$$
$$\overline{N'} = \overline{N} + 1 = 0\ 1\ 1\ 0$$
$$\overline{(1)\ 0\ 1\ 1\ 1}$$

Step 3. Ignore carry generated in step 2, confirming that result is +ve, and therefore the answer is 0111.

Case: (*b*) $M - N$ **where** $M < N$

Step 1. Find the 2's complement of N (larger).
Step 2. Add this to smaller number M. Here carry will not be there confirming that the result is −ve.
Step 3. Take 2's complement and put −ve sign.

Example 5.16 Subtract $(1010)_2$ from $(1000)_2$ using 2's complement method.

Solution $M - N = (1000)_2 - (1010)_2$

Step 1. 2's complement of N

$$\overline{N} = 0101$$
$$+ 1$$
$$\overline{N'} = \overline{N} + 1 = \overline{0110}$$

Step 2. Add result of step 1 to M

$$M = 1\ 0\ 0\ 0$$
$$\overline{N} + 1 = 0\ 1\ 1\ 0$$
$$\overline{1\ 1\ 1\ 0}$$

Step 3. Here no carry is generated, so result is −ve. Therefore, take 2's complement of it and put −ve sign i.e. $-(\overline{1110} + 1)$.

$$\therefore \text{1's comp.:} \quad 0\ 0\ 0\ 1$$
$$+ \ 1$$
$$\text{Negative Sign} \quad \overline{0\ 0\ 1\ 0}$$

$\therefore$ Result is $-(0010)_2$.

Summary of Subtraction by 1's and 2's Complement method for $(A-B)$ (For both cases $M > N$ or $M < N$)

By 1's complement	By 2's complement
• 1's complement of $B = \overline{B}$ = binary number with complement of each digit	2's complement of $B = \overline{B}' = \overline{B} + 1$
(a) Add $\overline{B}$ to A	(a) Add $\overline{B}'$ to A
(b) If carry is generated, result is +ve Add carry to the result Ans1: + $(A + \overline{B} + \text{carry})$ or If no carry generated take 1's complement of the result, then put −ve sign Ans2: $-\left(\overline{A+\overline{B}}\right)$	(b) If carry is generated, result is +ve Discard the carry Ans1: + $(A + \overline{B}')$ (carry discarded) or If carry is not generated take 2's complement of the result, then −ve sign Ans2: $-\left(\overline{A+\overline{B}'}\right)'$

5.4.3 9's Complement Subtraction

It is obtained by subtracting each digit in the number from 9.

Digit	0	1	2	3	4	5	6	7	8	9
9's complement	9	8	7	6	5	4	3	2	1	0

Case: (a) $M - N$ if $M > N$

Step 1. Take 9's complement of N (smaller).
Step 2. M + Step 1 carry will come to confirm that the result is +ve.
Step 3. Add carry to result. Given Ans.

Case: (b) $M - N$ if $M < N$

Step 1. Take 9's complement of N (larger).
Step 2. Add this number to M.
Step 3. Take 9's complement of result.

No carry indicates the result is negative.

Example 5.17 Subtract 84 from 93 using 9's complement.

Solution

Digit	0	1	2	3	4	5	6	7	8	9
9's complement	9	8	7	6	5	4	3	2	1	0

Step 1. $M - N = 93 - 84$
            Take 9's complement of $N = 15$.
Step 2. Add $M +$ Step 1.

M   9 3

9's complement of N +1 5
                      ―――――――
                       (1) 0 8

            Carry ――――↑
Step 3. Add carry to the result

                    0 8
      $\therefore$ + 1
                   ―――――
                    0 9   **Ans**.

Example 5.18 Subtract 38 from 20 using 9's complement.

i.e. $M - N = 20 - 38$.

Solution

Step 1. Take 9's complement of $N = 61$.
Step 2. Add M to Step 1.

            2 0
           +6 1
          ―――――
            8 1

        As no carry is there, therefore, result is −ve.
Step 3. 9's complement of above step 2.
            If no carry then negative $= -1$ 8 **Ans**.

5.4.4   10's Complement Subtraction

The 10's complement of the number can be found by adding 1's to 9's complement of the number.

Digit	0	1	2	3	4	5	6	7	8	9
9's complement	9	8	7	6	5	4	3	2	1	0

Case: (*a*) $M - N$ if $M > N$

Step 1. Take 10's complement of N (smaller) = 9's complement + 1.
Step 2. Add M with step 1.
Step 3. Drop the carry put +ve sign to the rest giving Ans. Carry will come, confirming that result is +ve.

Case: (*b*) $M - N$ if $M < N$

Step 1. Take 10's complement of N (larger) = 9's complement + 1.
Step 2. Add M with step 1. No carry will be there proving that result will be –ve.
Step 3. Take 10's complement of result and put negative sign to get Ans.

Example 5.19 Using 10's complement, subtract 3250 – 92532.

Solution $M - N = 3250 - 92532$

Step 1. Take 10's complement of N

$$= \text{9's complement} + 1$$
$$= 0\,7\,4\,6\,7$$
$$\underline{+\ 1}$$
$$0\,7\,4\,6\,8$$

Step 2. Add M with step 1.

$$\therefore \qquad M = \qquad 3\,2\,5\,0$$
$$\text{10's complement of } N = \underline{+\ 0\,7\,4\,6\,8}$$
$$(0)1\,0\,7\,1\,8$$

Step 3. No carry means negative number. Take 10's complement of step 2.

$$= 8\,9\,2\,8\,1$$

No carry means
negative number $\underline{+\ 1}$
$\therefore$ add 1, put –ve = $\quad -8\,9\,2\,8\,2$ **Ans.**

Summary of Subtraction by 9's and 10's Complement ($R = M - N$)

By 9's complement	By 10's complement
(a) M > N Calculate $R = (M + 9$'s complement of $N)$. This R will give a carry, showing that value of R is +ve. Just discard carry ∴ Ans: $R = +(M + 9's$ Complement of $N)$	*(a) M > N* Calculate $R = (M + 10$'s complement of $N)$. This R will give a carry. Use it to put a +ve sign in the result R and dicard it. ∴ Ans: $R = +(M + 10$'s Complement of $N)$
(b) M < N Calculate $R_1 = (M + 9$'s complement of $N)$. This R_1 will not give any carry, showing that result has to be $R = -$ve. Now take 9's complement of R_1 and put −ve sign. ∴ Ans: $R = -[9's$ Complement of $(M + 9$'s Complement of $N)]$	*(b) M < N* Calculate $R_1 = (M + 9$'s complement of $N)$. This R_1 will not give any carry, showing that result has to $R = -$ve. Now take 10's complement of R_1 and put −ve sign. ∴ Ans: $R = -[10's$ Complement of $(M + 10$'s Complement of $N)]$

5.5 Binary Codes

Binary codes are the mode of representation of any discrete element of information distinct among a group of quantities. They play an important role in digital computers. The codes must be in binary because computers can hold 1's and 0's only. Binary codes merely change the symbols, not the meaning of the elements of information that they represent. Some of the different binary codes for the decimal digits are

(*a*) Binary-coded decimal (BCD) code or 8421 code.
(*b*) Excess 3 code.
(*c*) Gray code.

5.5.1 *Binary-Coded Decimal System*

A very commonly used number system in the digital electronics field is the binary-coded decimal system (BCD). This system is different from the ordinary number system that **it expresses each decimal digit as a '4-bit nibble'**. The important factor to be considered in BCD system is that the highest value of BCD should be 4-bit nibble for decimal number 9. Table 5.2 represents equivalent BCD values for their decimal numbers. When using the BCD numbers system, it should be noted that 'all zeros must be retained, unlike a binary number where leading zeros can be dropped'.

The BCD number system is used when it is necessary to transfer decimal information into or out of digital machine. Examples of digital machines include calculator, digital clocks, digital voltmeter, frequency counters, etc.

Table 5.2 Binary-coded decimal (BCD) values

Decimal	Binary-coded decimal (8421)
0	0 0 0 0
1	0 0 0 1
2	0 0 1 0
3	0 0 1 1
4	0 1 0 0
5	0 1 0 1
6	0 1 1 0
7	0 1 1 1
8	1 0 0 0
9	1 0 0 1
10	0 0 0 1 0 0 0 0
11	0 0 0 1 0 0 0 1
12	0 0 0 1 0 0 1 0
13	0 0 0 1 0 0 1 1
↓	↓
128	0 0 0 1 0 0 1 0 1 0 0 0
129	0 0 0 1 0 0 1 0 1 0 0 1

Example 5.20 Convert the decimal number $(357)_{10}$ to a binary-coded decimal (BCD).

Solution

$3_{10} =$

2	3	–
2	1	1
	0	1

↑(LSB)
$= (0011)_2 = $ MSB of final digit

$5_{10} =$

2	5	–
2	2	1
2	1	0
	0	1

(LSB)
$= (0101)_2 = $ Middle term of final digit

$7_{10} =$

2	7	–
2	3	1
2	1	1
	0	1

(LSB)
$= (0111)_2 = $ LSB of final digit

∴ BCD equivalent of $(357)_{10} = (0011\ 0101\ 0111)_{BCD}$.

5.5.2 Gray Code

Gray code is a code which is used to represent the digital data, when it is converted from analog data. The advantage of the gray code over binary numbers is that only one bit in the code group changes, when going from one number to next. When same like bit is transferred, *i.e.* 0 to 0 or 1 to 1, then it gives low value (0). When unlike bit is transferred, *i.e.* 0 to 1 or 1 to 0, then it gives high value (1).

Mathematically for gray code conversion from one binary bits to another.

$$\text{Like bits} \rightarrow \begin{cases} 0 \oplus 0 \rightarrow \text{output is } 0 \\ 1 \oplus 1 \rightarrow \text{output is } 0 \end{cases}$$

$$\text{Unlike bits} \rightarrow \begin{cases} 0 \oplus 1 \rightarrow \text{output is } 1 \\ 1 \oplus 0 \rightarrow \text{output is } 1 \end{cases}$$

where symbols $\oplus$ represent exclusive–OR gate (EX–OR) which will be read in subsequent section of Logic Gates (Art 5.7).

Example 5.21 Convert $(1\,0\,0\,1\,0\,0)_2$ into gray code.

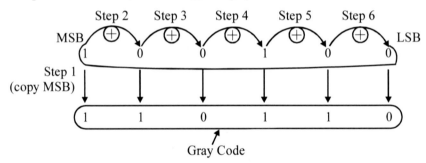

Solution

Step '1'. Write MSB number (*i.e.* 1) straight like that itself (*i.e.* copy).

Step '2'. Compare MSB number to next bit (*i.e.* 0).
$$1 \oplus 0$$
Unlike bits gives 1.

Step '3'. $0 \oplus 0$
Like bit gives 0.

Step '4'. $0 \oplus 1$
Unlike bit gives 1.

Step '5'. $1 \oplus 0$
Unlike bit gives 1.

Step '6'. $0 \oplus 0$
Like bit gives 0.

So gray code for $(100100)_2$ is $(110110)_{\text{gray}}$.

Example 5.22 Convert gray code 110110 into binary number.

Solution To convert gray code to binary number.

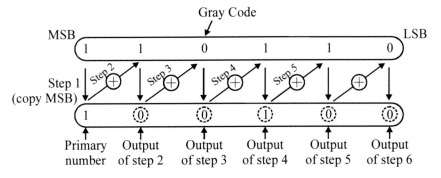

Step 1. Write MSB, as it is, *i.e.* 1 (*i.e.* just copy first bit).

Step 2. The output of step 1 is compared to next bit (1) gives 0, *i.e.*
$1 \oplus 1 \rightarrow 0$.

Step 3. The output of step 2 is compared to next bit (*i.e.* 0) gives, *i.e.*
$0 \oplus 0 \rightarrow 0$.

Step 4. The output of step 3 is compared to next bit (*i.e.* 1) gives, *i.e.*
$0 \oplus 1 \rightarrow 1$.

Step 5. The output of step 4 is compared to next bit (*i.e.* 1) gives, *i.e.*
$1 \oplus 1 \rightarrow 0$.

Step 6. The output of step 5 is compared to next bit, *i.e.* LSB (0) gives
$0 \oplus 0 \rightarrow 0$.

$\therefore$ Binary number for gray code $(110110)_{\text{gray}}$ is $(100100)_2$.

5.5.3 Excess 3 Code

It is another type of BCD code, which has 4-bit code, and is used with BCD numbers. To convert any decimal numbers into its **Excess 3** code form, add **3** to each decimal digit and then convert the sum to a BCD number. As weights are not assigned, it is kind of non-weighted code. Excess 3 is still used in cash registers and handheld portable electronic calculators.

Example 5.23 Convert the decimal number 8354 into Excess 3 code.

$$
\begin{array}{cccc}
8 & 3 & 5 & 4 \\
+\dfrac{3}{11} & \dfrac{3}{6} & \dfrac{3}{8} & \dfrac{3}{7} \quad \text{(add 3 to each digit)} \\
\downarrow & \downarrow & \downarrow & \downarrow \\
1011 & 0110 & 1000 & 0111 \quad \text{(Converted to 4 bit binary code)}
\end{array}
$$

Excess 3 code = (1011 0110 1000 0111)

5.6 Boolean Algebra and Its Properties

Boolean algebra is a set of rules, laws and theorems by which logical operations can be expressed mathematically. It was developed by George Boolean in the mid of eighteenth century. However, until 1938 the Boolean algebra was actually used in digital electronics.

Boolean algebra is a convenient, suitable and systematic way of expressing and analysing the operation of digital circuits and systems. In this algebra, variable can be either a zero or one. The binary digits are utilized to represent the two levels that occur within digital logic circuit. A binary bit '1' will represent a High level and a binary bit '0' will represent a Low level. The complement of a variable is represented by a 'bar' over the letter. For example, the complement of A is represented by $\overline{A}$.

5.6.1 Basic Laws of Boolean Algebra

Boolean Algebra follows the basic laws for Boolean addition and multiplications only. These are described below.

5.6.1.1   Boolean Addition (Logical 'OR')

Boolean addition involves variables having values of either a binary 1 or a binary 0. The basic rules of Boolean addition are as follows:

$0 + 0 = 0$ *i.e.* Low + Low = Low
$0 + 1 = 1$ *i.e.* Low + High = High   *i.e.*   High + Any No. = High
$1 + 0 = 1$ *i.e.* High + Low = High   1 + Any Binary digit = 1
$1 + 1 = 1$ *i.e.* High + High = High   1 + 1 + 1   = 1

The Boolean addition is same as the logical or 'OR' operation. The addition rules for Boolean algebra are the same as the binary addition rules as discussed in binary arithmetic (see Sect. 5.3).

5.6.1.2   Boolean Multiplication (Logical 'AND')

The multiplication rules for Boolean algebra are same as the binary multiplications as discussed in Sect. 5.3, *i.e.*

$$0 \cdot 0 = 0$$
$$0 \cdot 1 = 0$$
$$1 \cdot 0 = 0$$
$$1 \cdot 1 = 1$$

The Boolean multiplication is the same as the logical 'AND' operation.

5.6.2   Properties of Boolean Algebra

There are basically three properties of Boolean algebra. These are

(*a*) Commutative property,
(*b*) Associative property and
(*c*) Distributive property.

5.6.2.1   Commutative Property

The Boolean algebra is commutative when

$$A + B = B + A \tag{5.1a}$$

$$A \cdot B = B \cdot A \tag{5.1b}$$

The first equation (Eq. 5.1a) says that the order in which the variables are ORed makes no difference. The second equation (Eq. 5.1b) also states that the order in which the variables are ANDed makes no difference.

5.6.2.2 Associative Property

These properties are as follows:

$$A + (B + C) = (A + B) + C \tag{5.2a}$$

$$A \cdot (B \cdot C) = (A \cdot B) \cdot C \tag{5.2b}$$

Equation (5.2a) for associative properties of Boolean algebra states that it makes no difference in which order the variables are grouped when ORing several variables. Similarly, Eq. (5.2b) states that there will be no difference when ANDing several variables.

5.6.2.3 Distributive Property

The distributive property allows the factoring or multiplying out of expressions. Distributive properties are as follows:

$$A \cdot (B + C) = AB + AC, \tag{5.3a}$$

$$A + BC = (A + B)(A + C) \tag{5.3b}$$

$$A + \overline{A}B = A + B. \tag{5.3c}$$

Let us see the proof for Eq. (5.3b), *i.e.*

$$
\begin{aligned}
A + BC &= (A + B)(A + C) \\
RHS &= (A + B)(A + C) \\
&= A \cdot A + A \cdot C + A \cdot B + BC \\
&= A + AC + AB + BC \qquad (A \cdot A = A) \\
&= A(1 + C + B) + BC \\
&= A + BC \qquad (\because \; 1 + X = 1) \\
&= \text{LHS}
\end{aligned}
$$

Now let us see the proof for Eq. (5.3c), *i.e.*

$$A + \overline{A}B = A + B$$
$$\text{LHS} = A + \overline{A}B$$
$$= (A + \overline{A})(A + B) \quad \text{(using Eq. 5.3b properly)}$$
$$= 1(A + B)$$
$$= A + B$$
$$= \text{RHS}$$

5.6.3   De-Morgan's Theorem

There are two theorems as follows:

1. First theorem states that 'The complement of a sum equals the product of the complements'. It may be expressed as

$$\overline{A + B} = \overline{A} \cdot \overline{B} \tag{5.4a}$$

Similarly, $\overline{A + B + C + \cdots} = \overline{A} \cdot \overline{B} \cdot \overline{C} \cdots$.

2. De-Morgan's second theorem states that the 'Complement of the product equals sum of complement'.

$$\overline{A \cdot B} = \overline{A} + \overline{B} \tag{5.4b}$$

Similarly, $\overline{A \cdot B \cdot C \cdots} = \overline{A} + \overline{B} + \overline{C} + \cdots$.

Both the theorems can be proved with first-induction method.

A	B	$\overline{A}$	$\overline{B}$	$\overline{A+B}$	AB	$\overline{A} \cdot \overline{B}$	A+B	$\overline{A+B}$	$\overline{AB}$
0	0	1	1	1	0	1	0	1	1
0	1	1	0	1	0	1	0	1	1
1	0	0	1	1	0	1	0	1	1
1	1	0	0	0	0	0	1	0	0

$$\overline{A + B} = \overline{A} \cdot \overline{B} \ (\text{I}^{\text{st}} \text{ Theorem})$$

$$\overline{A \cdot B} = \overline{A} + \overline{B} \ (2^{\text{nd}} \text{ Theorem})$$

Some basic Boolean laws are listed in Table 5.3.

Table 5.3 List of Boolean laws

$A + 0 = A$	$A + \bar{A} = 1$
$A + 1 = 1$	$A \cdot \bar{A} = 0$
$1 +$ Any term or expression $= 1$	$(\bar{\bar{A}}) = A$
$A \cdot 0 = 0$	$A + BC = (A + B)(A + C)$
$A \cdot 1 = A$	$A + AB = A$
$A + A = A$	$A + \bar{A}B = (A + B)$
$A \cdot A = A$	$AB + \bar{B}C + BC = AB + C$

Example 5.24 Simplify the following Boolean algebra

$$(a)\ A + AB + A\bar{B}C \qquad (b)\ (\bar{A} + B)C + ABC$$
$$(c)\ \bar{A}B + AB + \bar{A}\bar{B}$$

Solution

$(a)\ A + AB + A\bar{B}C$

$$= A(1 + \underbrace{B + \bar{B}C}_{X}) \qquad\qquad (\because 1 + X = 1)$$

$$= A \qquad \textbf{Ans.}$$

$(b)\ (\bar{A} + B)C + ABC$

$$= (\bar{A} + B)C + ABC$$
$$= \bar{A}C + BC + ABC$$
$$= \bar{A}C + BC(1 + A) \qquad (\because\ 1 + A = 1)$$
$$= \bar{A}C + BC \qquad \textbf{Ans.}$$

$(c)\ \bar{A}B + AB + \bar{A} \cdot \bar{B}$

$$= \bar{A}(B + \bar{B}) + AB$$
$$= \bar{A}(B + \bar{B}) + AB$$
$$= \bar{A} \cdot 1 + AB \qquad\qquad (\because\ B + \bar{B} = 1)$$
$$= (\bar{A} + A)(\bar{A} + B) \qquad \text{(Distributive Property)}$$
$$= 1(\bar{A} + B) \qquad\qquad [A + BC = (A + B)(A + C)]$$
$$= \bar{A} + B \quad \textbf{Ans.}$$

Example 5.25 Simplify the following Boolean algebra

$$(a)\ AB + \bar{A}C + A\bar{B}C(AB + C) \qquad (b)\ Y = (\bar{A} + B)(A + B)$$
$$(c)\ Y = AB + \bar{A}C + A\bar{B}C(AB + C) \qquad (d)\ Y = (AB + \bar{C})(\overline{A + B + C})$$

Solution

(a) $AB + \overline{A}C + A\overline{B}C(AB + C)$

$$= AB + \overline{A}C + A\overline{B}C \cdot AB + A\overline{B}C \cdot C \qquad \begin{bmatrix} B\overline{B} = 0 \\ CC = C \\ AA = A \end{bmatrix}$$

$$= AB + \overline{A}C + 0 + A\overline{B}C$$

$$= A(B + \overline{B}C) + \overline{A}C \qquad \text{(Apply Distributive law)}$$

$$= A[(B + \overline{B})(B + C)] + \overline{A}C \qquad [B + \overline{B} = 1]$$

$$= A(B + C) + \overline{A}C$$

$$= AB + AC + \overline{A}C$$

$$= AB + C(A + \overline{A}) \qquad (A + \overline{A} = 1)$$

$$= AB + C \quad \textbf{Ans.}$$

(b) $Y = (\overline{A} + B)(A + B)$

$$= \overline{A} \cdot A + \overline{A}B + AB + B \cdot B$$

$$= 0 + \overline{A}B + AB + B$$

$$= B(\overline{A} + A + 1) \qquad (\because \ 1 + \text{Anything} = 1)$$

$$\therefore \ Y = B \quad \textbf{Ans.}$$

(c) $Y = AB + \overline{AC} + A\overline{B}C(AB + C)$

$$= AB + (\overline{A} + \overline{C}) + A\overline{B}C \cdot AB + A\overline{B}C \cdot C \quad [\text{Apply De-Morgan's Theorem}]$$

$$= AB + \overline{A} + \overline{C} + 0 + A\overline{B}C \qquad \because \ \overline{B} \cdot B = 0$$

$$= A(B + \overline{B}C) + \overline{A} + \overline{C} \qquad [\text{Apply Distributive law}]$$

$$= A[(B + \overline{B})(B + C)] + \overline{A} + \overline{C}$$

$$= A(B + C) + \overline{A} + \overline{C}$$

$$= AB + AC + \overline{A} + \overline{C}$$

$$= (\overline{A} + AB) + (\overline{C} + AC)$$

$$= (\overline{A} + A)(\overline{A} + B) + (\overline{C} + A)(\overline{C} + C)$$

$$= 1(\overline{A} + B) + (\overline{C} + A) \cdot 1$$

$$= \overline{A} + B + A + \overline{C}$$

$$= (A + \overline{A}) + B + \overline{C}$$

$$= 1 + B + \overline{C} \qquad \because \ [A + \overline{A} = 1]; [1 + B = 1]; [1 + \overline{C} = 1]$$

$$\therefore Y = 1 \quad \textbf{Ans.}$$

(d) $Y = \overline{(AB + \overline{C})(\overline{A} + B + C)}$

$\quad Y = \overline{(AB + \overline{C})(\overline{A} \cdot \overline{B} + C)}$ [Apply De-Morgan's Theorem]

$\quad Y = \overline{[AB\overline{A}.B + ABC + \overline{A} \cdot \overline{B} \cdot \overline{C} + C\overline{C}]}$ $\because [A\overline{A} = 0; C\overline{C} = 0, BB = 1]$

$\quad Y = \overline{ABC + \overline{A} \cdot \overline{B} \cdot \overline{C}}$

$\quad Y = \overline{ABC} \cdot \overline{\overline{A} \cdot \overline{B} \cdot \overline{C}}$ [Apply De-Morgan's Theorem]

$\quad Y = (\overline{A} + \overline{B} + \overline{C})(A + B + C)$ **Ans**.

Example 5.26 Simplify the following Boolean expression

   (a) $Y = \overline{A}C[\overline{\overline{ABD}}] + \overline{A}BCD + A\overline{B}C$
   (b) $X = \overline{A}B + ABD = \overline{AB}C\overline{D} + BC$

Solution

(a) $Y = \overline{A}C[\overline{\overline{ABD}}] + \overline{A}BCD + A\overline{B}C$

$\quad Y = \overline{A}C[\overline{\overline{A}} + \overline{B} + \overline{D}] + \overline{A}BCD + A\overline{B}C$ [De − Morgan's Theorem]

$\quad Y = \overline{A}C[A + \overline{B} + \overline{D}] + \overline{A}BCD + A\overline{B}C$

$\quad Y = 0 + \overline{A}\overline{B}C + \overline{A}C\overline{D} + \overline{A}BCD + A\overline{B}C$ $\because (A \cdot \overline{A} = 0)$

$\quad Y = \overline{B}C(A + \overline{A}) + \overline{A}D(C + B\overline{C})$ $\because \;\; [A + \overline{A} = 1]$

$\quad Y = \overline{B}C + \overline{A}D(C + B)(C + \overline{C})$ [distributive property]

$\quad Y = \overline{B}C + \overline{A}D(C + B)$ $[C + \overline{C} = 1]$

$\quad Y = \overline{B}C + \overline{A}DC + \overline{A}DB$ **Ans**.

(b) $X = \overline{A}B + ABD + \overline{A}BC\overline{D} + BC$

$\quad = B(\overline{A} + AD) + \overline{C}(\overline{AB}D + B)$

$\quad = B(\overline{A} + A)(\overline{A} + D) + C(B + \overline{AD}) \cdot (B + \overline{B})$ [Distributive properly]

$\quad = B(\overline{A} + D) + C(B + \overline{AD})$ $\begin{bmatrix} A + \overline{A} = 1 \\ B + \overline{B} = 1 \end{bmatrix}$

$\quad = \overline{A}B + BD + BC + \overline{AD}C$ **Ans**.

Example 5.27 Using Boolean algebra, prove that

   (a) $(A + B)(\overline{AC} + C)(\overline{B} + AC) = \overline{A}B$
   (b) $(A + B)(\overline{AC} + C)(\overline{B} + AC) = \overline{A}B$
   (c) $(A + B)(\overline{A} + C) = AC + \overline{A}B + BC$

Solution

(a) $LHS = (A+B)(\overline{A}\ \overline{C}+C)(\overline{\overline{B}+AC})$

$= (A\overline{A}\ \overline{C}+AC+B\overline{A}\ \overline{C}+BC)(\overline{\overline{B}}\cdot\overline{AC})$ [De-Morgan's theorem]

$= (0+AC+\overline{A}B\overline{C}+BC)(B\cdot\overline{AC})$ $\begin{bmatrix}\overline{\overline{B}}=B\\ A\cdot\overline{A}=0\end{bmatrix}$

$= (AC+\overline{A}B\overline{C}+BC)[B(\overline{A}+\overline{C})]$

$= (AC+\overline{A}B\overline{C}+BC)(\overline{A}B+B\overline{C})$

Applying De-Morgan's Theorem:

$LHS = AC\cdot\overline{A}B+AC\cdot B\overline{C}+\overline{A}B\overline{C}\cdot\overline{A}B+\overline{A}B\overline{C}\cdot B\overline{C}$

$\quad\quad +BC\cdot\overline{A}B+BC\cdot B\overline{C}$

$= 0+0+\overline{A}B\overline{C}+\overline{A}B\overline{C}+\overline{A}BC+0$ $\begin{bmatrix}A\cdot\overline{A}=0\\ C\cdot\overline{C}=0\\ A\cdot A=A\end{bmatrix}$

$= \overline{A}B\overline{C}+\overline{A}BC$ $[X+X=X]$

$= \overline{A}B(\overline{C}+C)$

$= \overline{A}B\cdot 1$

$= \overline{A}B = RHS$

∴   LHS = RHS

(b) $LHS = (A+B)(B+C)(C+A)$

$= [AB+AC+B\cdot B+BC][C+A]$

$= [AB+AC+B+BC][C+A] \Rightarrow [B\cdot B=B]$

$= (AB+AC+B(1+C))(C+A)$

$= (AB+AC+B)(C+A)$ $[1+C=1]$

$= [B(1+A)+AC](C+A)$

$= (B+AC)(C+A)$ $[1+A=1]$

$= BC+AB+AC+AC$ $[C\cdot C=C]$

$= AB+BC+AC$

∴   LHS = RHS

(c) $LHS = (A+B)(\overline{A}+C)$

$= A\overline{A}+AC+\overline{A}B+BC$ $(A\cdot\overline{A}=0)$

$= 0+AC+\overline{A}B+BC$

$= AC+\overline{A}B+BC$

∴   LHS = RHS

5.7   Logic Gates

Logic gates are the basic building blocks of digital systems. A logic gate is a digital circuit with one or more inputs but with one output. The relationship between input and output is based on 'certain stage', so its known as logic gates. Logic gates are used to implement the different logic functions based on the Boolean algebra.

Classification of Gates

Basic Gates
1. OR Gate
2. AND Gate
3. NOT (Invertor) Gate

Universal Gates
1. NAND Gate
2. NOR Gate

Special Purpose
1. EX-OR Gate
2. EX-NOR Gate

5.7.1   OR Gate (A + B)

The OR gate performs logical addition, commonly known as OR function. The OR gate has two or more inputs and only one output.

Logical operation

The operation of OR gate is such that a high (1) on the output is produced when any of the inputs is high (1). The output is low (0) only when all the inputs are low (0).

As shown in Fig. 5.1a, A and B represent the inputs and Y the output. The resistance R is the load resistance.

*   If $A = 0$ and $B = 0$, then $V_o = 0$ and $Y = 0$.
*   If $A = 1$ and $B = 0$, diode D_1 will conduct, so the output $Y = 1$.
*   If $A = 0$ and $B = 1$, diode D_2 will conduct, the output $V_o \neq 0$ and $Y = 1$.
*   If $A = 1$ and $B = 1$, both diodes will conduct, so the output $Y = 1$.

The electrical equivalent circuit of an OR gate is shown in Fig. 5.1b where switches A and B are connected in parallel with each other. If either A, B or both are

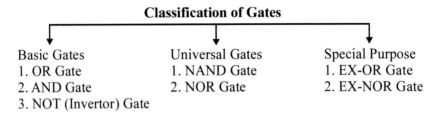

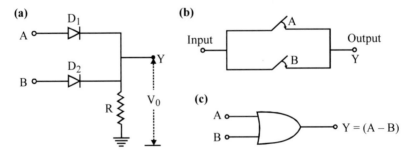

Fig. 5.1 a OR gate no bias required here, **b** electrical equivalent circuit and **c** logic symbol

Input		Output
A	B	Y = A + B
0	0	0
0	1	1
1	0	1
1	1	1

Table 5.4 Truth table of an OR gate

closed then only the output will come. The logic symbol for OR gate is shown in Fig. 5.1c.

The logic operation of the two-input OR gate is described in the truth table shown in Table 5.4. In fact it is A + B = logical addition.

5.7.2 AND Gate (A.B)

The AND gate performs logical multiplication, commonly known as AND function. The AND gate is composed of two or more inputs and a single output.

Logical operation: The output of AND gate is high only when all the inputs are high. When any of the inputs is low, the output is low.

As shown in Fig. 5.2a, A and B represent the inputs and Y represents the output.

* If $A = 0$ and $B = 0$, both diodes conduct. As they are forward biased, so the $V_0 = 0$ output $Y = 0$.
* If $A = 0$ and $B = 1$, D_1 conducts and D_2 does not conduct, so the output $V_0 = 0$ and $Y = 0$.
* If $A = 1$ and $B = 0$, D_1 does not conduct and D_2 conducts, so the output $V_0 = 0$ and $Y = 0$.
* If $A = 1$ and $B = 1$, both diodes do not conduct as they get reverse biased, $V_0 \neq 0$ and so the output $Y = 1$.

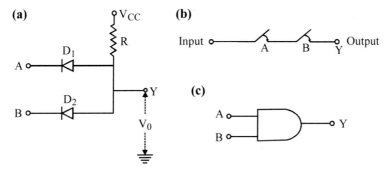

Fig. 5.2 a AND gate note the bias V_{CC}, **b** electrical equivalent circuit and **c** logic symbol

Table 5.5 Truth table of an AND gate

Input		Output
A	B	$Y = A \cdot B$
0	0	0
0	1	0
1	0	0
1	1	1

The electrical equivalent circuit of an AND gate is shown in Fig. 5.2b, where two switches A and B are connected in series. If both A and B are closed, then only output will come.

Logical symbols of the AND gate is shown in Fig. 5.2c. The logic operation of the two-input AND gates is described in the truth table shown in Table 5.5. In fact $A \cdot B$ = Logical multiplication.

5.7.3 NOT Gate (Inverter) $(\overline{A})$

The NOT gate performs a basic logic function called inversion or complementation. The purpose of gate is to change one logic level to opposite level. It has one input and one output.

Logical operation: When a high level is applied to an inverter input, a low level will appear at its output and vice versa.

As shown in Fig. 5.3a, A represents the input and Y represents the output.

* If the input is high, the transistor is in ON state and the output is low, V_0 ≈ '0' = Y.
* If the input is low, the transistor is in OFF state and the output is high, V_0 = '1' = Y.

The logic symbol for the inverter is shown in Fig. 5.3b.

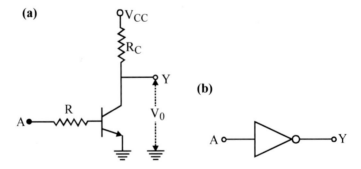

Fig. 5.3 a NOT gate and **b** logic symbol

Table 5.6 Truth table of a NOT gate

Input	Output
A	$Y = \bar{A}$
0	1
1	0

The truth table is given in Table 5.6.

5.7.4   Universal Gates

With the help of three basic gates, *i.e.* OR, AND and NOR gates, two other gates can be formed. These are called the NAND and NOR gates. NAND and NOR gates are also known as universal gates because they can be used individually as the universal building blocks to implement any logic circuits. Also any Boolean expression can be implemented either by use of NAND gates only or by NOR gates only. Therefore, these gates have become very popular and are extensively used in digital circuitry. This is because in IC fabrication repeating a circuit on the same water is much easy, and therefore mass fabrication makes it cheaper.

5.7.4.1   NAND Gate ($\overline{A \cdot B}$)

NAND is an abbreviation of NOT-AND. It has two or more inputs and only one output.

Logical operation: When all the inputs are high, the output is low. However, if any of the inputs is low, the output is high. The logic symbol for the NAND gate is shown in Fig. 5.4a.

The truth table for the NAND gate is shown in Table 5.7.

Fig. 5.4   a Logic symbol of NAND gate and **b** Logic symbol of NOR gate

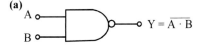

Table 5.7 Truth table of a NAND gate

Input		AND output	NAND output
A	B	Y = A · B	$Y = \overline{A \cdot B}$
0	0	0	1
0	1	0	1
1	0	0	1
1	1	1	0

Table 5.8 Truth table of a NOR gate

Input		OR output	NOR output
A	B	A + B	$Y = \overline{A + B}$
0	0	0	1
0	1	1	0
1	0	1	0
1	1	1	0

5.7.4.2 NOR Gate $(\overline{A + B})$

NOR is the abbreviation of NOT-OR. It has two or more inputs but only one output.

Logical operation: When any of the inputs is high, the output is low. Only when all the inputs are low, the output is high. The logical symbol of NOR gate is shown in Fig. 5.4b.

Truth table for the NOR gate is shown in Table 5.8.

5.7.5 Special Purpose Gates: EX-OR $(A \oplus B)$ and EX-NOR $(A \odot B)$ Gates

Exclusive OR gate or EXOR or XOR gate and exclusive NOR gate or EX-NOR or XNOR gate are two special types of gates.

(a) **EXOR Gate**: It is a gate with two or more inputs and one output. The output of a two-input EXOR gate assumes a HIGH state if one and only one input assumes a HIGH state. This is equivalent to state that the output is HIGH if either input A or input B is HIGH exclusively, and low when both are 1 or 0 simultaneously. Therefore, output is L, when both A and B are different.

The logic symbol for EX-OR gate is shown in Fig. 5.5a and truth table for the EX-OR operation is given in Table 5.9.

The truth table of EX-OR gates shows that the output is HIGH when anyone input, but not all of the inputs, is at 1. These exclusive features eliminate a similarity to the OR gate. The EX-OR gate responds with a High output only when an ODD

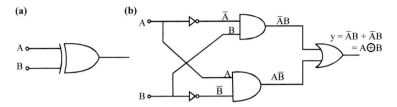

Fig. 5.5 a Logic symbol of EX-OR gate and **b** EX-OR gate using NOT-AND-OR gate (basic gates)

Table 5.9 Truth table of a two-input EXOR gate

Input		Output
A	B	$Y = A \oplus B = A\bar{B} + \bar{A}B$
0	0	0
0	1	1
1	0	1
1	1	0

number of inputs are HIGH. When there is an even number of HIGH inputs, such as two or four, then the output will always be low.

The main characteristic properties of an EXOR gate is that it can perform modulo-2 addition or half adder. The name half adder or module-2 adder refers to the fact that possible carry resulting from addition of the preceding bits has not been taken into account.

Half adder

A	B	Sum $S = A \oplus B$	Carry $C = AB$
0	0	0	0
0	1	1	0
1	0	1	0
1	1	0	1

Here the sum $S = A \oplus B = A\bar{B} + \bar{A}B$.

Another important property of an EXOR gate is that it can be used as a controlled inverter, *i.e.* by using EX-OR gate, a logic variable can be complemented or allowed to pass through it unchanged. This is done by using one EXOR input as shown in Fig. 5.6. When the control input is HIGH, the output is $Y = \bar{A}(\text{low})$ and its vice versa is also true.

Fig. 5.6 EXOR gate as a controlled inverter

Fig. 5.7 EX-NOR gate

Table 5.10 Truth table of two-input EX-NOR gate

Input		Output
A	B	$Y = \overline{A \oplus B} = A \odot B = AB + \overline{A}\,\overline{B}$
0	0	1
0	1	0
1	0	0
1	1	1

(b) **EX-NOR gate**: It is an EX-OR gate followed by an inverter. An EX-NOR gate has two or more inputs and one output. The output of two-input EX-NOR gate assumes a High state if all the inputs are HIGH or if all the inputs are LOW. If any one of the two inputs is HIGH or LOW, then the output is always LOW. Hence, EX-NOR can be used for bit comparator. The output of an EX-NOR gate is 1 if both the inputs are similar, *i.e.* both are 0 or 1, otherwise its output is 0. It can also be used in coincidence circuit and even parity checker or odd parity checker.

The logic symbol of EX-NOR gate is shown in Fig. 5.7 and truth table is shown in Table 5.10.

The Boolean expression for the EX-NOR is

$$Y = \overline{A \oplus B}$$
$$= A\overline{B} + \overline{A}B$$
$$= (\overline{A} + B)(A + \overline{B})$$
$$Y = AB + \overline{A}\,\overline{B} = A \odot B$$

5.7.6   *Implementation of Basic Gates and Special Purpose Gates Using Minimum Number of NAND and NOR Gate*

Minimum NAND gates required	OR	AND	NOT	EXOR	EX-NOR	NOR	NAND
	3	2	1	4	5	4	1
Minimum NOR gates required	2	3	1	5	4	1	4

Example 5.28 Implement OR, AND and NOT gate with universal gate, *i.e.* (1) NAND gates only (2) NOR gates only.

Solution (1) With NAND gates only:

(a) OR Gate $(Y = A + B)$

$$Y = \overline{\overline{A} \cdot \overline{B}} = A + B$$

(b) AND Gate $(Y = AB)$

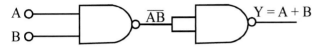

$$Y = A + B$$

(c) NOT gate $(Y = \overline{A})$

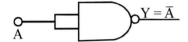

$$Y = \overline{A}$$

(d) NOR by NAND

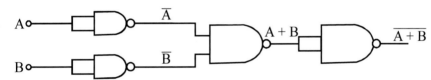

$$A + B \qquad \overline{A + B}$$

(2) With NOR gates only:

(a) OR Gate

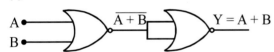

$$\overline{A + B} \qquad Y = A + B$$

(b) AND Gate

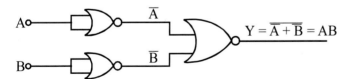

$$Y = \overline{\overline{A} + \overline{B}} = AB$$

(c) NOT Gate

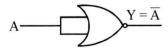

$$Y = \overline{A}$$

(d) NAND by NOR

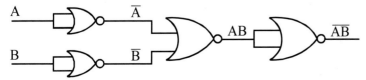

$$AB \qquad \overline{AB}$$

Example 5.29 Implement EXOR and EX-NOR with minimum
  1. NAND Gate 2. NOR Gate

Solution (1) (a) EXOR gate with NAND gate

Minimum number of NAND Gate = 4 (refer Sect. 5.7.6).

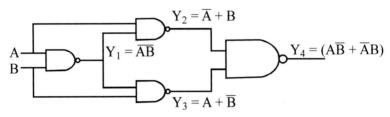

$$Y_1 = \overline{A \cdot B}$$

$$Y_2 = \overline{Y_1 \cdot A} = \overline{\overline{A \cdot B} \cdot A} = \overline{\overline{A \cdot B}} + \overline{A}$$

(De-Morgan theorem)

$$= AB + \overline{A} \qquad\qquad \text{(Distributive Property)}$$

$$= (\overline{A} + B)(\overline{A} + A) \qquad \therefore (\overline{A} + A = 1)$$

$$= \overline{A} + B$$

$$Y_3 = \overline{Y_1 \cdot B} = \overline{\overline{AB} \cdot B} \qquad\qquad \text{(De-Morgan theorem)}$$

$$= \overline{\overline{A \cdot B}} + \overline{B}$$

$$= AB + \overline{B} \qquad\qquad \text{(Distributive property)}$$

$$= (A + \overline{B})(B + \overline{B}) \qquad \therefore (B + \overline{B} = 1)$$

$$= A + \overline{B}$$

$$Y_4 = \overline{Y_2 \cdot Y_3} = \overline{(\overline{A} + B)(A + \overline{B})}$$

$$= \overline{A\overline{A} + \overline{A}\,\overline{B} + AB + B \cdot \overline{B}} \qquad \therefore (A \cdot \overline{A} = 0) \text{ and } (B \cdot \overline{B} = 0)$$

$$= \overline{\overline{A}\,\overline{B} + AB}$$

$$= \overline{\overline{A}\,\overline{B}} \cdot \overline{AB} \qquad\qquad \text{(De-Morgan's Theorem)}$$

$$= (\overline{\overline{A}} + \overline{\overline{B}})(\overline{A} + \overline{B})$$

$$= (A + B)(\overline{A} + \overline{B})$$

$$= A\overline{B} + \overline{A}B = A \oplus B$$

(b) EX-NOR gate with NAND gate

Minimum number of NAND gate = 5 (refer Sect. 5.7.6).

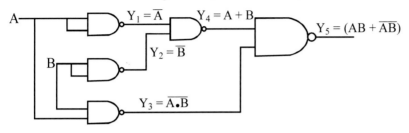

$$Y_1 = \overline{A \cdot A} = \overline{A}$$
$$Y_2 = \overline{B \cdot B} = \overline{B}$$
$$Y_3 = \overline{A \cdot B}$$
$$Y_4 = \overline{Y_1 \cdot Y_2} = \overline{\overline{A} \cdot \overline{B}} \qquad \text{(De-Morgan's Theorem)}$$
$$= \overline{\overline{A}} + \overline{\overline{B}}$$
$$= A + B$$
$$Y_5 = \overline{Y_3 \cdot Y_4} = \overline{\overline{A \cdot B} \cdot (A + B)}$$
$$= \overline{\overline{A \cdot B}} + \overline{(A + B)}$$
$$= A \cdot B + \overline{AB}$$
$$= A \odot B$$

(2) (a) EXOR gate with NOR gate

Minimum number of NOR gate required = 5 (see Sect. 5.7.6).

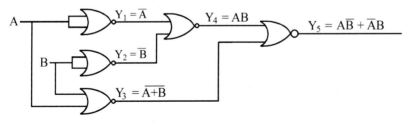

$$Y_1 = \overline{A+A} = \overline{A}$$
$$Y_2 = \overline{B+B} = \overline{B}$$
$$Y_3 = \overline{A+B}$$
$$Y_4 = \overline{\overline{A}+\overline{B}} \qquad\qquad\qquad \text{(De-Morgan's theorem)}$$
$$= \overline{\overline{A}} \cdot \overline{\overline{B}}$$
$$= A \cdot B$$
$$Y_5 = \overline{Y_3+Y_4} = \overline{\overline{A+B}+A\cdot B} \qquad \text{(De-Morgan's Theorem)}$$
$$= \overline{\overline{A+B}} \cdot \overline{AB}$$
$$= (A+B)(\overline{AB}) = (A+B)(\overline{A}+\overline{B})$$
$$= A\overline{A} + A\overline{B} + \overline{A}B + B\overline{B}$$
$$= A\overline{B} + \overline{A}B = A \oplus B$$

(b) EX-NOR gate with NOR gate
Minimum number of NOR gate required = 4.

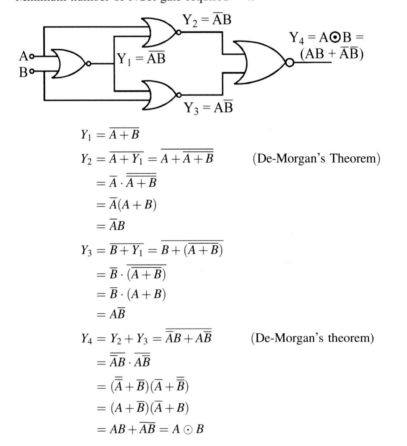

$$Y_1 = \overline{A+B}$$
$$Y_2 = \overline{A+Y_1} = \overline{A+\overline{A+B}} \qquad \text{(De-Morgan's Theorem)}$$
$$= \overline{A} \cdot \overline{\overline{A+B}}$$
$$= \overline{A}(A+B)$$
$$= \overline{A}B$$
$$Y_3 = \overline{B+Y_1} = \overline{B+(\overline{A+B})}$$
$$= \overline{B} \cdot \overline{(\overline{A+B})}$$
$$= \overline{B} \cdot (A+B)$$
$$= A\overline{B}$$
$$Y_4 = \overline{Y_2+Y_3} = \overline{\overline{A}B+A\overline{B}} \qquad \text{(De-Morgan's theorem)}$$
$$= \overline{\overline{A}B} \cdot \overline{A\overline{B}}$$
$$= (\overline{\overline{A}}+\overline{B})(\overline{A}+\overline{\overline{B}})$$
$$= (A+\overline{B})(\overline{A}+B)$$
$$= AB + \overline{AB} = A \odot B$$

5.8   Important Terms from Boolean Function

5.8.1   Minterms

A binary variable may appear either in its normal form (A) or in its complement form $(\overline{A})$. Now consider two binary variables A and B combined with an AND operation. Since each variable may appear in either form, there are four possible (2^2 = 4) combinations:

$$\overline{A}\ \overline{B}, \overline{A}B, A\overline{B} \text{ and } AB$$

Each of the four AND terms represents one of the distinct areas in the Venn diagram as shown in Fig. 5.8 and is called a '*minterm*', (*or a 'standard product'*). In a similar manner, '*n*' variables can be combined to form 2^n minterms. The symbol of the minterm is 'm_j', where '*j*' denotes the decimal equivalent of the binary number of the minterm designated.

5.8.2   Maxterms

n variables forming an OR term, with each variables being primed or unprimed, providing 2^n possible configurations. These are called '*maxterms*' (*or 'standard sums*'). The symbol of **Maxterm is 'M_j'**.

Minterms and Maxterms for three binary variables are as follows:

S. N.	A	B	C	Minterm terms	Designation m_j minterm	Maxterm terms	Designation M_j maxterm
1	0	0	0	$\overline{A}\ \overline{B}\ \overline{C}$	m_0	$A+B+C$	M_0
2	0	0	1	$\overline{A}\ \overline{B}\ C$	m_1	$A+B+\overline{C}$	M_1
3	0	1	0	$\overline{A}\ B\ \overline{C}$	m_2	$A+\overline{B}+C$	M_2
4	0	1	1	$\overline{A}\ B\ C$	m_3	$A+\overline{B}+\overline{C}$	M_3
5	1	0	0	$A\ \overline{B}\ \overline{C}$	m_4	$\overline{A}+B+C$	M_4
6	1	0	1	$A\ \overline{B}\ C$	m_5	$\overline{A}+B+\overline{C}$	M_5
7	1	1	0	$A\ B\ \overline{C}$	m_6	$\overline{A}+\overline{B}+C$	M_6
8	1	1	1	$A\ B\ C$	m_7	$\overline{A}+\overline{B}+\overline{C}$	M_7

(Note $8 = 2^3 = 2^n$ for n = variable)

Fig. 5.8 Venn diagram of two variables for minterms

$\overline{A}\overline{B}, \overline{A}B, A\overline{B}$ and AB

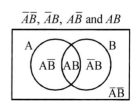

Min and MaxTerms of Two Variables

A	B	Minterm	Maxterms
0	0	$\overline{A}\,\overline{B} = m_0$	$A + B = M_0$
0	0	$\overline{A}B = m_1$	$A + \overline{B} = M_1$
0	1	$A\overline{B} = m_2$	$\overline{A} + B = M_3$
0	1	$AB = m_3$	$\overline{A} + \overline{B} = M_4$

Minterms and Maxterms of a given number ABC

If ABC = 011; Minterm of ABC = $\overline{A}\,BC = 011$ and its Maxterm = $A + \overline{B} + \overline{C} = 1 + 0 + 0$

5.8.3   Canonical Forms

Boolean functions expressed as a sum of minterms or product of maxterms are said to be in canonical form.

By '*sum of minterms*' means ORing of terms.

For example, for three binary variables (see Table above).

$F = m_1 + m_4 + m_6 + m_7$ can be written in short notation as

$$\text{Sum of products (minterms)} = F(A, B, C) = \Sigma(1, 4, 6, 7)$$

Here the summation symbol Σ stands for the ORing of terms while the number following it is the minterm of the function.

By *product of maxterms* means ANDing of terms, for example, for three binary variables (see Table above).

$F = M_0\, M_3\, M_5\, M_5\, M_7$ can be written in short notation as

$$\text{Product of sum (maxterms)} = F(A, B, C) = \Pi(0, 3, 5, 6, 7)$$

The product symbol, Π, denotes the ANDing of maxterms; the numbers are the maxterms of the function.

Note: Canonical form means all the variables should be there in each term, for example, $(A + B)(B + C)$ or $(ABC + A + BC)$ are not in canonical form but $(\overline{A} + B + C)(\overline{A} + B + C)$ or $(\overline{A}BC + A\overline{B}C)$ are in canonical form.

5.8.4   Standard Form of Boolean Expressions

In this configuration, the terms that form the function may contain one, two or any number of literals.

There are two types of standard forms.

1. Sum of product (SOP)
2. Product of sum (POS)

5.8.5   Sum of Product (SOP)

The sum of products is a Boolean expression containing AND terms, called *product terms* of one or more literals each. The *sum* denotes the ORing of these terms. An example of function expressed in sum of the product is

$$F_1 = \overline{A} + C + \overline{A}BC$$

Here the expression has three product terms of one, two and three literals each, respectively. Their sum is in effect an OR operation.

5.8.6   Product of Sums (POS)

The product of sums is a Boolean expression containing OR terms, called *sum terms*. Each term may have any number of literals. The *product* denotes the ANDing of these terms. An example of a function expressed in product of sum is

$$F_2 = A(B + \overline{C})(A + \overline{B} + C + \overline{D})$$

This expression has three terms of one, two or four literals each, respectively. The product is an AND operation.

5.8.7   Conversion Between Two Canonical Forms

To convert from one canonical form to another, interchange the symbol Σ and Π and list those numbers which are missing from the original form. In order to find the missing terms, one must realize that the **total number of minterms or maxterms is 2^n, where n is the number of binary variables** in the function.

For example, if

$$F(A, B, C) = \Sigma(1, 2, 5, 6, 7)$$

then conversed form is

$$F(A, B, C) = \Pi(0, 3, 4)$$

Example 5.31 Obtain the following Boolean expression in a canonical SOP form.

$$(a)\ Y(A, B) = A + B \quad (b)\ Y(A, B, C) = A + BC$$
$$(c)\ Y = AB + ACD$$

Solution

(a) $Y(A, B) = A + B$

$$= A \cdot 1 + B \cdot 1$$

(Both variables are required in the product)

$$= A \cdot (B + \overline{B}) + B \cdot (A + \overline{A})\ \text{As}\ [A + \overline{A} = 1; B + \overline{B} = 1]$$
$$= AB + A\overline{B} + AB + \overline{A}B$$
$$= AB + A\overline{B} + \overline{A}B$$
$$= m_3 + m_2 + m_1 \qquad\qquad\qquad\qquad \ldots\text{Canonical SOP form}$$
$$\therefore \ \ Y(A, B) = \Sigma(1, 2, 3)$$

(b) $Y(A, B, C) = A + BC$

$$= A \cdot 1 \cdot 1 + BC \cdot 1$$

(each term should have all three variable product)

$$= A(B + \overline{B})(C + \overline{C}) + BC(A + \overline{A})$$
$$= A(BC + B\overline{C} + \overline{B}C + \overline{B}\,\overline{C}) + ABC + \overline{A}BC$$
$$= ABC + AB\overline{C} + A\overline{B}C + A\overline{B}\,\overline{C} + \overline{A}BC \rightarrow \text{std. form}$$
$$\therefore \ \ Y(A, B, C) = (m_7 + m_6 + m_5 + m_4 + m_3) = \Sigma(3, 4, 5, 6, 7)\ \ \ldots\text{Canonical form}$$

(c) $Y(A, B, C, D) = AB + ACD$

$$Y = AB \cdot 1 \cdot 1 + ACD \cdot 1 \quad \text{(All four variable required)}$$
$$Y = AB(C + \overline{C})(D + \overline{D}) + ACD(B + \overline{B})$$
$$Y = AB(CD + C\overline{D} + \overline{C}D + \overline{C}\,\overline{D}) + ACDB + ACD\overline{B}$$
$$Y = ABCD + ABC\overline{D} + AB\overline{C}D + AB\overline{C}\,\overline{D} + ABCD + A\overline{B}CD$$
$$Y = ABCD + ABC\overline{D} + AB\overline{C}D + AB\overline{C}\,\overline{D} + ABCD + A\overline{B}CD$$

Std. form (Total possible terms $= 2^n = 2^4 = 16$)
$$Y = (m_{15} + m_{14} + m_{13} + m_{12} + m_{11}) = \Sigma(11, 12, 13, 14, 15) \rightarrow \text{Canonical form}$$

Example 5.32 Obtain the following Boolean expression in a canonical POS form:

$$(a)\ Y = (\overline{A} + \overline{B})(B + C)(A + \overline{C}) \quad (b)\ Y = A + BC$$

Solution

(a) $Y = (\overline{A} + \overline{B})(B + C)(A + \overline{C})$

    (One variable missing in each sum)

$Y = (\overline{A} + \overline{B} + 0)(B + C + 0)(A + \overline{C} + 0)$

$Y = (\overline{A} + \overline{B} + C \cdot \overline{C})(B + C + A \cdot \overline{A})(A + \overline{C} + B \cdot \overline{B})$

$Y = (\overline{A} + \overline{B} + C)(\overline{A} + \overline{B} + \overline{C})(B + C + A)$

$(B + C + \overline{A})(A + \overline{C} + B)(A + \overline{C} + \overline{B})$ [Distributive property]

$Y = M_6 \cdot M_7 \cdot M_0 \cdot M_1 \cdot M_2 \cdot M_3$

$Y = \pi(0, 1, 2, 3, 6, 7) \rightarrow$ Canonical form

(b) $Y = A + BC$ ...(SOP to POS conversion)

    Apply Distribution Property

$Y = (A + B)(A + C)$ (Making it POS)

$Y = (A + B + 0)(A + 0 + C)$

    (One variable missing in each sum therefore we add that variable)

$Y = (A + B + C\overline{C})(A + B\overline{B} + C)$

$Y = (A + B + C)(A + B + \overline{C})(A + B + C)(A + \overline{B} + C)$

$Y = (A + B + C)(A + B + \overline{C})(A + \overline{B} + C)$

$Y = M_0 M_1 M_2$

$Y = \Pi(0, 1, 2)$

Example 5.33 Express the Boolean function $Y = A + \overline{B}C$ in

  (a) Canonical sop   (b) Canonical POS

Solution

(a) $Y = A + \overline{B}C$

$= A \cdot 1 \cdot 1 + \overline{B}C \cdot 1 \quad [A + \overline{A} = B + \overline{B} = C + \overline{C} = 1]$

$= A(B + \overline{B})(C + \overline{C}) + \overline{B}C(A + \overline{A})$

$= A(BC + B\overline{C} + \overline{B}C + \overline{B}\overline{C}) + A\overline{B}C + \overline{A}\overline{B}C$

$= ABC + AB\overline{C} + A\overline{B}C + A\overline{B}\overline{C} + A\overline{B}C + \overline{A}\overline{B}C$

$= ABC + AB\overline{C} + A\overline{B}C + A\overline{B}\overline{C} + \overline{A}\overline{B}C$ Standard form

$Y = m_7 + m_6 + m_5 + m_4 + m_1$

$Y = \Sigma(1, 4, 5, 6, 7) \rightarrow$ canonical SOP

(b) $Y = A + \overline{B}C$ (Apply Distributive Property)

$\quad Y = (A + \overline{B})(A + C)$

$\quad Y = (A + \overline{B} + 0)(A + C + 0)$

$\quad Y = (A + \overline{B} + C \cdot \overline{C})(A + C + B \cdot \overline{B}) \text{As}(B \cdot \overline{B} = C \cdot \overline{C} = 0)$

$\quad Y = (A + \overline{B} + C)(A + \overline{B} + \overline{C})(A + B + C)(A + \overline{B} + C)$

$\quad\quad = (A + \overline{B} + C)(A + \overline{B} + \overline{C})(A + B + C)$ Standard POS

$\quad Y = M_2 \cdot M_3 \cdot M_0$

$\quad Y = \Pi(0, 2, 3) \rightarrow$ Canonical POS

5.9   Karnaugh Map or K-Map: Representation of Logical Functions

Karnaugh map (or K-map) technique provides a systematic method for simplifying and manipulating Boolean expressions. In this technique, the information contained in a truth table or available in SOP or POS form is represented on K-map. In an n-variable K-map, there are 2^n cells. Each cell corresponds to one of the combinations of n variables. Therefore, we see that for each row of the truth table, *i.e.* for each minterm and for each maxterm, there is one specific cell in the K-map. The variables have been designated as A, B, C and D and the binary numbers formed by them are taken as AB, ABC and ABCD for two, three and four variables, respectively. The K-map for two, three and four variables is shown in Fig. 5.9a, b, c, with serial number of cells.

The cell number in the K-map signifies the serial number in the box are of increasing value of the binary (see Table 5.1). **The numbering is so done that two adjacent cells will have a change in one variable only,** *i.e.* A or B or C or D, so as to follow the gray code. Two changes in adjacent cells are not permitted.

In the actual use of K-map, the cell numbers are not written but the value of Y, the Boolean expression of that cell, is written.

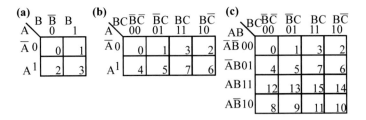

Fig. 5.9 K-maps: **a** two variable, **b** three variable and **c** four variable (sequence as per gray code not binary). Cell number in the boxes are values of the corresponding binary number

5.9.1 Pairs, Quads and Octets of K-Map

K-maps are used to simplify the logic circuits. But before using them we must have an idea about pairs, quads and octets.

(a) **Pairs**: Let us consider K-map shown in Fig. 5.10, which contains terms formed by horizontally and vertically adjacent pair. The first one represents the product $AB\overline{C}\,\overline{D}$ and the second one represents the products $A\,\overline{B}\,\overline{C}\,\overline{D}$. It shows that as we go from the first '1' to the second '1,' only one variable, *i.e.* B goes from uncomplemented to complemented. The other variables A and D do not change their form. Whenever this happens, we can eliminate the variable that changes its form.

Hence, the algebraic equation:

$$Y = AB\overline{C}\,\overline{D} + A\,\overline{B}\,\overline{C}\,\overline{D} \quad \text{can be written directly as}$$
$$Y = A\overline{C}\,\overline{D}$$

(b) **The Quad**: When four adjacent boxes of terms are grouped horizontally, vertically or in a square, as shown in Fig. 5.11, the group so formed is called as quad.

(c) **The Octet**: When eight adjacent terms in boxes are grouped horizontally or vertically, as shown in Fig. 5.12, the group so formed is called an octet. In other words, eight adjacent is equal to two adjacent squares or quads.

5.9.2 Steps to Simplify Logic Circuits by K-Map

Karnaugh map is an important tool to simplify logic circuits. The steps to be followed are given below:

1. Enter a 1 on K-map for each fundamental product that corresponds to 1 output in the truth table. Enter 0 s at all other places but this is optional.
2. Encircle the possible octets, quads and pairs to get the largest group possible. Never forget to roll the map and to overlap 1's wherever required.

Fig. 5.10 K-maps with pairs of variable

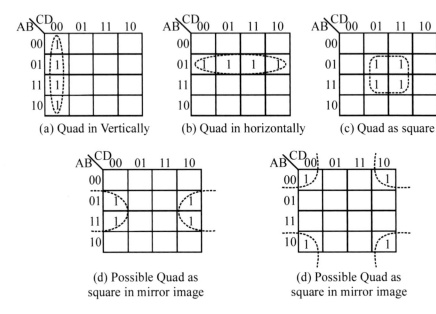

(a) Quad in Vertically   (b) Quad in horizontally   (c) Quad as square

(d) Possible Quad as
square in mirror image

(d) Possible Quad as
square in mirror image

Fig. 5.11 Possible quad forming in K-map

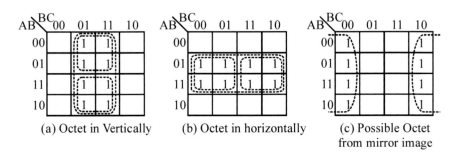

(a) Octet in Vertically   (b) Octet in horizontally   (c) Possible Octet
from mirror image

Fig. 5.12 Possible octet forming in K-map

3. Always reject the redundant grouping.
4. Encircle the isolated 1's if left otherwise.
5. Write the Boolean expression or equation by ORing the products corresponding to the encircled groups.
6. Draw to logic circuit for the Boolean equation.

Questions

Fill up the blanks:

1. Give the decimal value of binary 10010.
2. The output of an AND gate is LOW when _____.
3. The output of a NOT gate is HIGH when _____.
4. The output of an OR gate is LOW when _____.
5. The fractional binary number 0000.1010 in decimal is _____.
6. The fractional decimal number 6.75 in binary is _____.
7. The decimal value of binary 10000110 is _____.
8. The decimal values of 2^3 and 2^{-1} are_____ and _____, respectively.
9. The fractional binary number 10010.0100 in decimal _____.
10. The output of an AND gate with three inputs, A, B and C, are HIGH when _____.
11. What is the binary equivalent of the decimal number 368.
12. The simplification of the Boolean expression $(\overline{\overline{AB}\overline{C}}) + \overline{(A\overline{B}C)}$ is _____.
13. The 2's complement of the number 1101101 is _____.
14. The equivalent of $(3A.C)_{16} = (......)_2 = (........)_8$.
15. De-Morgan's first theorem shows the equivalence of

    (a) NOR gate and bubbled AND gate.
    (b) NOR gate and NAND gate.
    (c) OR gate and exclusive OR gate.
    (d) NAND gate and NOT gate.

16. The hexadecimal number 'A0' has the decimal value equivalent to _____.
17. The decimal value of $(AC)_{16} + (110)_2 + (765)_8$ is _____.

Short Questions

1. If a three-input NOR gate has eight input possibilities, how many of those possibilities will result in a HIGH output?
2. Write short notes on

    (a) Binary number system.
    (b) Decimal number system.
    (c) Octal number system.
    (d) Hexadecimal number system.

3. Convert the following decimal numbers to the indicated bases:

    (a) 7562.45 to octal.
    (b) 1938.257 to hexadecimal.
    (c) 175.175 to binary.

4. Convert the following numbers from the given base to the other three bases indicated:

    (a) Decimal 225 to binary, octal and hexadecimal.
    (b) Binary 11010111 to decimal, octal and hexadecimal.
    (c) Octal 6223 to decimal, binary and octal.
    (d) Hexadecimal 2AC5 to decimal, binary and octal.

5. Write the procedure to obtain 1's and 2's complement of a binary number.
6. Write the procedure to obtain gray code from binary number
7. Write the procedure to obtain binary number from gray code.
8. Prove with Boolean algebra $(x + y)(x + z)$ simplifies to $x + yz$.
9. Find the 1's and 2's complement of the following eight-digit binary numbers:

    (a) 10100111,
    (b) 10000001 and
    (c) 00000000.

10. What is Boolean algebra?
11. State De-Morgan's theorem.
12. What are the three basic logic gates? Give its truth tables.
13. Subtract 110111 from 1111001 with 1's and 10's complement methods.
14. What are universal gates? Why are they called so? Give its truth tables.
15. Subtract 11011001 from 11110010 with 2's complement method.
16. Draw the logic diagram corresponding to the following Boolean expressions without simplifying them:

    (a) $Y = ABC + A'B + BC$,
    (b) $Y = ABC(D + E'F)$ and
    (c) $Y = (A + B)(C + D)(A' + B + D)$.

Long Questions

1. Explain the operation of NOT, NOR, EX-OR and NAND gates with truth table.
2. State and prove De-Morgan's theorem.
3. Show that the solution is same for (1010-1101) with 1's, 2's, 9's and 10's complement method.
4. Do the following:

    (a) Find the decimal equivalent of $(1011.011)_2$.
    (b) Convert $(276)_{10}$ to octal.
    (c) Convert $(2D6)_{16}$ to binary.

(d) Find the binary equivalent of $(48.442)_{10}$.

(e) Find the binary addition $(1101 + 1111)_2$.

5. Realize the logic expression $Y = \overline{B}\,\overline{C} + \overline{A}\,\overline{C} + \overline{A}\,\overline{B}$, using basic gates and with only NAND gates.

6. Implement the following expressions using logic gates:

(a) $Y = A + BC\overline{D}$ using NAND gates.

(b) $Y = (A + C)(A + \overline{D})(A + B + \overline{C})$ using NOR gates.

7. Explain how BCD addition is carried out

8. Minimize the following Boolean expression:

(a) ABCD + ABD,

(b) A(A + B),

(c) $XY + XYZ + XY\overline{Z} + \overline{X}YZ$ and

(d) $AB + (\overline{A}\,\overline{C}) + A\overline{B}C(AB + \overline{C})$.

9. Do the following:

(a) Convert the given expression in standard SOP form: $f(A, B, C) = A + ABC$.

(b) Convert the given expression in standard POS form. $f(A, B, C) = A \cdot (A + B + C)$.

(c) Convert the gray code (1011011011110l010) to binary number.

(d) Convert the binary code (1010101101111) to gray code.

(e) Give an example of Excess 3 code.

Chapter 6
Transducers

Contents

6.1 Introduction

A transducer is a device that converts energy from one form to another form. This energy may be electrical, mechanical, chemical, optical or thermal. They are basically sensor of various types used in instruments that measures / displays temperature, strain, pressure, force, torque, acceleration, vibration, liquid or gas flow, viscosity, humanity, light intensity, etc.

Transducer may be classified according to their application, method of energy conversion, nature of the output signal and so on. All these classifications usually result in overlapping areas. A sharp distinction among the types of transducers is difficult.

© Springer Nature Singapore Pte Ltd. 2020

S. S. Srikant and P. K. Chaturvedi, *Basic Electronics Engineering*,

https://doi.org/10.1007/978-981-13-7414-2_6

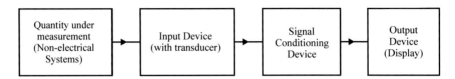

Fig. 6.1 A basic instrumentation system

The transducer that gives electrical energy as output is called electrical transducer. The output electrical signal may be voltage, current, of any frequency and production of these signals is based upon resistive, capacitive, inductive effects, etc. For measuring non-electrical quantities, a detector is used which usually converts the physical quantity into a displacement that activates the electrical transducers. An instrumentation system generally consists of three components, namely, an input device, a signal conditioning device and an output device as shown in Fig. 6.1. The input device with transducer in its, senses the quantity under measurement and changes it to a proportional electrical signal. In the signal conditioning device, the signal is amplified, filtered or otherwise modified to a level acceptable to a storage device or display or output device.

The input measurement quantities for most instrumentation systems are non-electrical. The non-electrical quantity is converted to an electrical signal by a device called transducer and is used for measurement and control. In general, the energy transmission by a transducer may be mechanical, electrical, magnetic, thermal, radiant or chemical.

6.2 Transducer Classification—Passive and Active

There are two basic types of transducers, namely, Active transducers and Passive transducers

Active Transducer
Such transducers converts energy (mostly non-electrical energy) directly from one form to another, without the need for our external electrical/power source or excitation.

Passive Transducer
Such transducers are also called as externally powered transducers. In these transducers, the power is required for energy conversion from an external source. However, they may also absorb some energy from the physical phenomenon under study. In passive transducers, some of the parameters other than electrical parameter change with light, pressure, power, etc.

A few examples of active and passive transducers are shown in Table 6.1.

Table 6.1 Active and passive transducers

Active transducer	Passive transducers
• Thermocouple	• Resistance
• Piezoelectric transducer	– Potentiometric device
• Photoelectric transducer	– Resistance strain gange
	– Thermistor
	– Photoconductive cell
	• Inductive
	– Linear variable differential transformer
	• Capacitance
	• Voltmeter and ammeter
	• Hall effect device

6.3 Basic Requirements of Transducer

Some of the basic requirements of a transducer are listed below:

1. **Range**: The transducer must be operable over the minimum to maximum values of the parameters to be measured.
2. **Accuracy**: It is in conformity with the accepted value. A transducer should be accurate.
3. **Repeatability**: The closeness of the output value, among a number of consecutive measurements for the same value of the input, is called repeatability. A transducer should have better repeatability.
4. **Sensitivity**: The ratio of change in output to the change in input should be high enough for the transducer for better resolution of the system.
5. **Environmental effects**: The influence of temperature, acceleration, shock, vibration and corrosion should not affect the performance of transducer.
6. **Noise**: The output signal from the transducer should not be affected by noise.
7. **Loading effect**: It is determined by the mass, exterior size, geometric, configuration of the transducer. All transducers absorb some energy because of these factors. So the loading effect should be less.
8. **Output impedance**: This should be matched with the rest of the measuring systems.
9. **Power requirements**: Proper required voltage should be applied for the externally excited transducers.
10. **Frequency response**: The transducer system must have an accurate response to the maximum rate of change of the system. The change of output with input frequency should be flat over the measurement range.
11. **Linearity**: The input and output characteristics of a transducer should be linear.
12. **Ruggedness**: The transducer should be rugged enough to withstand the over-loads.

6.4   Passive Transducers

The definition and various types of passive transducers are already discussed under Sect. 6.2 and in Table 6.1. The details of some of the important passive transducers are given below.

6.4.1   Resistive Transducers

Here a resistive element of a material is given by

$$R = \frac{\rho L}{A} \text{ ohms}$$

where

R is the resistance in ohms,
ρ is the specific resistance in ohm-metre,
l is the length of the material in metres and
A is the area of cross section in square metres.

By causing change in displacement (linear or angular), pressure (strain) and temperature, we can change the values of 'l', 'A' and 'p', respectively. Potentiometer resistance, gauge, resistance strain gauge, resistance thermometer and thermistor are some of the examples of resistance transducers.

6.4.2   Potentiometer (POT)

It is a passive device with a thin wire of platinum or nickel alloy of 0.01 mm diameter carefully wound on an insulated former. The resistance element can be used by either *dc* or *ac* voltage supply. It has a sliding contact called movable slider or wiper, whose motion is either linear or rotary. The potentiometer for both linear motion and rotary motion is shown in Fig. 6.2. The potentiometer called pot in short has carbon layered bakelite ring on which wiper 'c' rotates.

Here V_i = input voltage (in volt),
V_0 = output voltage (in volt),
l_t = total length (in meter),
l_i = displacement of wiper (in metres) and
θ_i = angular displacement (in degrees).

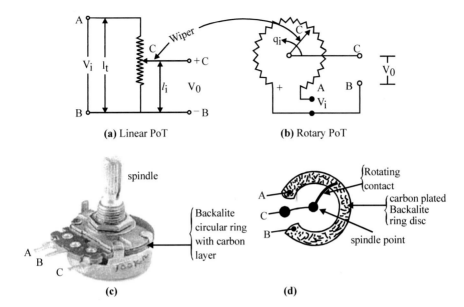

Fig. 6.2 Potentiometer: (**a**) linear pot, (**b**) rotary pot, (**c**) photo of a real pot and (**d**) linear carbon plated resistance ring inside a pot

6.4.3 Electrical Strain Gauge

If a metal conductor is stretched or compressed, its resistance changes because of dimensional changes (length and cross-sectional area) and resistivity change. If a wire is under tension and increase its length from l to $l + \Delta l$, i.e. the strain $S = \Delta l/l$, increases, then its resistance also increases from R to $R + \Delta R$.

The sensitivity of a strain gauge is described in terms of characteristics called the gauge factor G, defined as the unit change in resistance per unit change in length, i.e.

$$G = \frac{\Delta R/R}{\Delta l/l} = \frac{\Delta R/R}{S} \tag{6.1}$$

When a gauge wire is subjected to a positive strain, its length increases whereas its area of cross section decreases. Since the resistance is a function of length and area of cross section, it also changes. The resistance wire strain gauges are classified into two forms:

1. Unbonded strain gauge and
2. Bonded strain gauge.

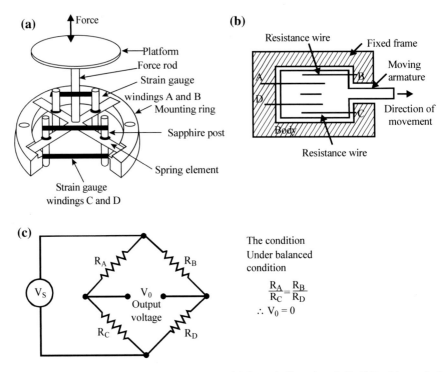

Fig. 6.3 (**a**) Star spring unbonded strain gauge, with four winding wires A, B, C, D; (**b**) stretched unbonded strain gauge, with four wires A, B, C, D; (**c**) bridge circuit for unbonded strain gauge under balanced condition, $V_0 = 0$

6.4.3.1   Unbonded Strain Gauge

It consists of four parts mounted on a star spring structure as shown in Fig. 6.3a. The four sapphire posts hold four equal lengths of tungsten–platinum wire of 5 μm diameter of resistances R_A, R_B, R_C and R_D. The temperature compensation is obtained by the Wheatstone bridge circuit formed by these four wires. When the star spring is subjected to the force under measurement, the strain gauge elements on the opposite sides of the spring gets strained in opposite polarity. The unbalanced voltage of the bridge circuit is proportional to pressure (due to the applied force) on the displacement.

6.4.3.2   Bonded Strain Gauge

There are four types of bonded strain gauge. They are (a) linear strain gauge, (b) rosette, (c) torque gauge and (d) helical gauge as shown in Fig. 6.4. The bonded strain gauge consists of a single fine resistance wire made of nichrome or constantan or nickel or platinum in the form of a grid of 25 μm diameter cemented to a

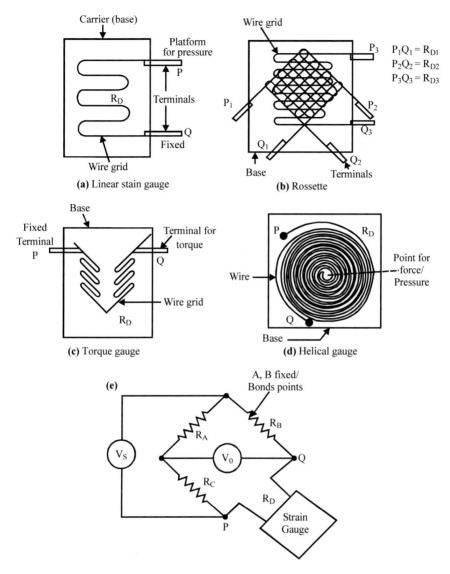

Fig. 6.4 Bonded strain gauges

carrier made of thin paper or bakelite or Teflon sheets. The spreading of wire ensures uniform distribution of stress. The carrier is bonded to the structure under test by means of an adhesive materials like epoxy, cement or bakelite cement or ethyl-cellulose cement, etc. If the strain gauge is stretched or compressed, the resistance of the conductor changes. This change in resistance is proportional to the applied strain and is measured using bridge circuit. Whenever the bridge circuit balances, no current flows through the meter, then the meter reads zero ($V_0 = 0$)

(Figs. 6.3e). When strain gauge is subjected to strain, resistance of the strain gauge changes, affecting the balance of the circuit. Hence, the meter shows some reading which is proportional to applied strain. Dummy strain gauge is used to cancel the temperature effect to the active gauge. When the direction of strain is unknown, then three resistances R_{D1}, R_{D2}, R_{D3} are used called rosette strain gauge. The bridge decides the direction of the strain by which resistance has changed R_{D1} or R_{D2} or R_{D3}.

6.4.4   Resistance Thermometer

It is a primary electrical transducer used for measuring temperature changes in terms of resistance changes. A metal or an alloy is used as an resistive element. The specific resistance of the metal increases with increase in temperature. As the temperature changes, the length and specific resistance change, and hence the change in resistance. Platinum, nickel, copper and tungsten are the commonly used metals for the wire which are wound on mica or ceramic formers as shown in Fig. 6.5a. The fluid whose temperature is to be measured is brought in direct contact with open element. The drawback with this type is that (a) if the fluid is in motion, it will cause error in measurement (b) if the fluid is corrosive it will spoil the element. This could be avoided by enclosing the elements in a protective tube of Pyrex or porcelain or quartz or nickel (Fig. 6.5b).

The resistance of most electrical conductors varies with temperature according to the relation

$$R = R_0(1 + \alpha T + \beta T^2 + \ldots) \tag{6.2}$$

where

R_0 resistance at temperature T_0 (usually 0 °C),
R resistance at T and
α, β constants and T is the rise in temperature above T_0.

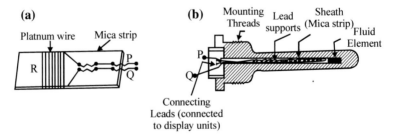

Fig. 6.5 General purpose resistance thermometer (RT): **a** wire-wound RT; **b** coil wire in a protective enclosure

Over a small temperature range, depending on the material, the above equation is linear and reduces to

$$R = R_0(1 + \alpha T)$$
$$\therefore \quad T = (R - R_0)/(\alpha R_0) \tag{6.3}$$

where α is the temperature coefficient of resistance.

Important properties of materials used for resistance thermometer are: (i) high-temperature coefficient of resistance; (ii) stable properties, so that the resistance characteristic does not drift with repeated heating and cooling or mechanical strain; and (iii) a high resistivity to permit the construction of small sensors.

The variation of resistivity with temperature up to 800 °C of some of the materials used for resistance thermometers is shown in Fig. 6.6. From this figure, it is observed that tungsten has a suitable temperature coefficient of resistance but is brittle and difficult to form. Copper has a low resistivity and is generally confined to applications where the sensor size is not restricted. Both platinum and nickel are widely used because they are relatively easy to obtain in the pure state. Platinum has an advantage over nickel in that:

(a) Its temperature coefficient of resistance is linear over a wide range of temperature from 0 °C to 800 °C:
(b) Very less drift of resistance with time.
(c) Suitable to precision application and accuracy.
(d) Do not get corroded by chemicals or atmosphere easily.

Fig. 6.6 Variation of resistivity with temperature of materials used for resistance thermometers

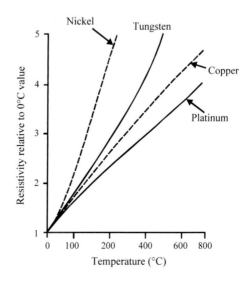

6.4.5 Thermistor

Thermistor or thermal resistor is a two-terminal semiconductor device, whose resistance is temperature sensitive. The value of such resistors decreases with increase in temperature. Materials employed in the manufacture of the thermistors include oxides of cobalt, nickel, copper, iron, uranium and manganese. It can be used for measuring temperature of −100 °C to 400 °C where the resistance changes from 1 to 10Ω (Fig. 6.7b). The thermistor has very high temperature coefficient of resistance of the order of 3–5% per °C, making it an ideal temperature transducer. Being a semiconductor material, the temperature coefficient of resistance is normally negative. The resistance at any temperature T is given approximately by

$$R_T = R_0 \exp\left[\beta\left(\frac{1}{T} - \frac{1}{T_0}\right)\right] \tag{6.5}$$

R_T thermistor resistance at temperature $T(K)$,
R_0 thermistor resistance at temperature $T_0(K)$ and
β a constant determined by calibration.

At high temperatures, Eq. (6.5) reduces to

$$R_T = R_0 \exp(\beta/T) \tag{6.6}$$

The resistance-temperature characteristic of a typical thermistor is shown in Fig. 6.7b. The curve is nonlinear and the drop in resistance from 5000 to 100Ω occurs for an increase from 20 °C to 100 °C.

The temperature of the device can be changed internally or externally. An increase in current through the device will raise its temperature, causing a drop into its terminal resistance. Any externally applied heat source will result in an increase in its body temperature and drop in resistance.

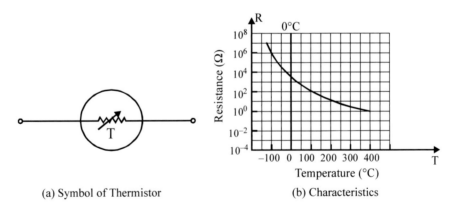

(a) Symbol of Thermistor   (b) Characteristics

Fig. 6.7 a Symbol and **b** resistance-temperature characteristics of a typical thermistor

Three useful parameters for characterizing the thermistor are the time constant, dissipation constant and resistance ratio. The time constant is the time for a thermistor to change its resistance by 63% of its initial value, for zero-power dissipation. Typical values of time constant range from 1 to 50 s.

The dissipation factor is the power necessary to increase the temperature of thermistor by 1 °C. Typical values of dissipation factor range from 1 to 10 m W/°C.

Resistance ratio is the ratio of the resistance at 25 °C to that at 125 °C. Its range is approximately 3–60.

Thermistors are used to measure temperature, flow, pressure, liquid level voltage or power level, vacuum, composition of gases and thermal conductivity and also in compensation network.

6.4.6 Capacitive Transducer

It is used to measure mechanical displacement without causing any additional mechanical loading on the vibrating object. The most commonly used capacitive transducer is a parallel plate capacitor without variable gap or variable overlapping areas 'A'.

The capacitance of a parallel plate is given by

$$C = \frac{\varepsilon_0 \varepsilon_r A}{d} \text{ Farads} \tag{6.7}$$

where

C   is the capacitance in farads,
ε_0 is the 8.854×10^{-12} farads/metre,
ε_r is the relative permittivity,
A   is the area of each plate in metre^2 and
d   is the distance between the plates in metres.

From Eq. 6.7, it is clear that capacitance is inversely proportional to the spacing between the parallel plates and is directly proportional to the plate area and permittivity of the dielectric medium. Thus, a variable capacitance can be made by varying 'd' or 'A' or 'ε_r' (Fig. 6.8).

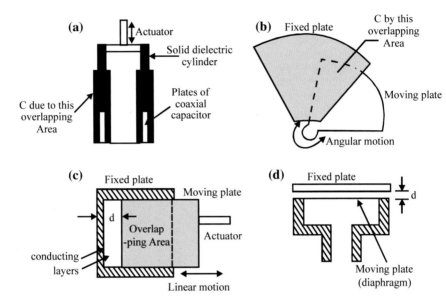

Fig. 6.8 Capacitive transducer with **a** variable dielectric, **b** variable plate area and **c** variable space

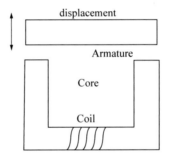

Fig. 6.9 Inductive transducer

6.4.7   *Inductive Transducer*

It is widely used to measure quantities such as pressure, acceleration, force and displacement or position. The inductance of a coil depends upon the magnetic flux linking with it. A change in the inductance is proportional to the quantity to be measured and is obtained by changing the flux linkage. A bridge or an oscillatory circuit can be used to measure the inductance. The variation of inductive (flux linkages) can be obtained by the movement of an armature as shown in Fig. 6.9.

6.4.8 Linear Variable Differential Transducer (LVDT)

It consists of a primary coil, two identical secondary coils and a rod-shaped magnetic core positioned centrally inside the coil. The displacement to be measured is transferred to the magnetic core the through suitable linkages. The two secondary windings have equally number of turns with current in opposite directions due to connection as in Fig. 6.10. When the core is symmetrically placed with respect to the two secondary coils, equal voltages but out of phase are induced in the two coils. Therefore, the output resultant differential voltage becomes zero (Fig. 6.10).

$$\therefore V_{S1} = -V_{S2}; V_{S1} + V_{S2} = 0 \text{ [when core is symmetrically placed]}.$$

This balance point is called null position. When the core gets displaced from the central position due to the displacement of the object linked mechanically, then $V_{S1} \neq -V_{S2}$. This results in differential output from the transformer, which is proportional to the displacement of the core and hence the object to be measured.

As shown in Fig. 6.10, LVDT is widely used in electronic comparators, thickness measuring units, level indicators, crop testing machines and numerically controlled machines. It has fine resolution, good stability and high accuracy.

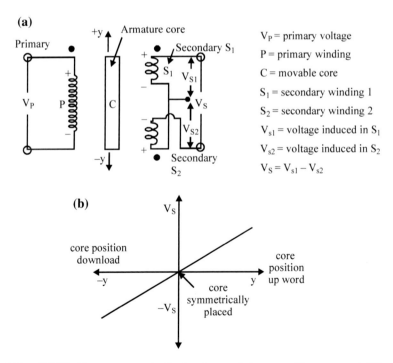

Fig. 6.10 **a** LVDT with core symmetrically placed at $y = 0$; **b** core position versus net differential voltage (Vs) in the secondary coil

6.4.9 Hall Effect

When on a current (I_x) carrying semiconductor of rectangular structure, a transverse magnetic field (B_z) is applied, then a voltage V_y is generated in it, which is perpendicular to both I_x and B_z. This phenomenon is called Hall effect and the voltage V_y as Hall voltage $V_y = V_H$.

In fact, this voltage generation is due to the charge moving in the magnetic field, experiencing magnetic Lorentz force $F_y = e(V_x \times B_z)$, by both electrons and holes in one direction (upward) only. (This is because electrons have velocity as well as charge both –ve while in holes both +ve) (Fig. 8.11). This is similar to the reason of Fleming's left-hand rule on a conductor carrying current (I_x) placed in a magnetic field (B_z) perpendicular to it, where the whole of conductor experiences upward (F_y) force.

As a result for p-type semiconductor, holes being the majority, the tops surface AA (Fig. 8.11) becomes +ve giving voltage (V_H) across AB as +ve. Similarly for n-type, voltage (V_H) across AB will be –ve. This charge accumulation on AA surface give rise to an electric field, known as Hall electric field (E_{Hy}), from AA to BB direction. This field in turn opposes further drift of charges due to magnetic field till it reaches equilibrium, when net Lorentz force (electric and magnetic force) = 0, $i.e.$

$$e \cdot E_{Hy} + e(v_x \times B_z) = 0$$
$$\therefore E_{Hy} = -(v_x \times B_z) = -v_x \cdot B_z \quad \text{(for } v_x \text{ and } B_z \text{ being perpendicular)} \tag{6.7}$$

Here we defined Hall coefficient R_H as the ratio of $E_{Hy}/(J_x \cdot B_z)$, where J_x = current density = $n\,e\,v_x$

$$\therefore |R_H| = |v_x/J_x| = +1/(ne) \tag{6.8}$$

Thus by sensing the sign of V_H as –ve or +ve, we can identify between n-type and p-type semiconductors. Also by Eq. 6.7, if the width of electric field (E_{Hy}) regions is 'd' then:

$$V_{Hy} = E_{Hy} \cdot d = v_x \cdot B_z = \frac{J_H}{ne} \cdot B_z \tag{6.8a}$$

Thus from Eq. (6.8a), if any three of the parameters are known, fourth one can be computed. Therefore, carrier doping density or magnetic field strength, can be known (Fig. 6.11).

Applications: The Hall effect transducers is prominent and widely used in semiconductor materials. The advantage of Hall effect transducers is that they are non-contact devices with high resolution and small size. Some of other applications are in measurement of velocity, rpm, sorting, liquid sensing, non-current and magnetic field measurement.

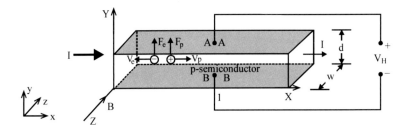

Fig. 6.11 Schematic arrangement to observe the Hall effect by current flow I

6.5   Active Transducer

The definition and various types of active transducers are already discussed under Sect. 6.2 and Table 6.1. The details of some of the importance active transducers are given below.

6.5.1   Thermocouples

A thermocouple consists of a pair of dissimilar metal wires joined together. At one end of the joint forms the **sensing or hot junction**. The other end of the dissimilar metal joint usually known as **reference or cold junction**, is maintained at a known constant temperature (reference temperature). When a temperature difference exists between the sensing junction and the reference junction, an emf is produced. The magnitude of this voltage depends on the material used for the wires and the temperature difference between the two junctions. When a junction is terminated by a voltmeter or recording instrument, as shown in Fig. 6.12, the meter indication is proportional to the temperature difference between the hot junction and the reference junction. This thermoelectric effect, caused by contact potentials at the junctions, is called the 'Seebeck effect', after the German physicist Thomas Seebeck.

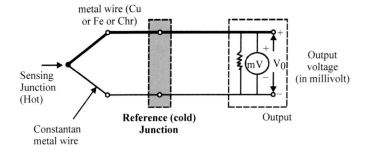

Fig. 6.12 A typical thermocouple circuit

Fig. 6.13 Thermo-emf for some common thermocouple material

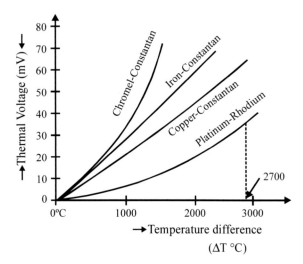

Thermocouples are made from a number of different metals pairs including copper–constantan, iron–constantan, chromel–constantan, platinum–rhodium, etc., as shown in Fig. 6.13. They cover wide range of temperature, going as high as 2700 °C. The values shown are based on reference temperature of 0 °C.

Advantages of the Thermocouple

(a) Rugged in construction.
(b) Covers wide range of temperature, from –270 °C to 2700 °C.
(c) Extension leads and compensating cables, can be used for a distant point temperature measurement. It is best suitable for temperature measurement of industrial furnaces.
(d) Response speed is high.
(e) Measurement accuracy is quite satisfactory.
(f) Calibration can easily be checked.
(g) Cheaper in cost.

Limitation of Thermocouple

(a) For accurate temperature measurements, cold junction compensation is needed.
(b) The emf induced versus temperature characteristics are somewhat nonlinear.
(c) Stray pickup voltage due to noise may be possible.
(d) In many applications, amplification of signal is needed.

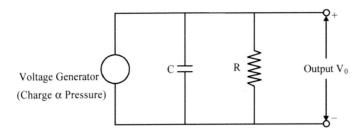

Fig. 6.14 Equivalent circuit of piezoelectric transducer

6.5.2   Piezoelectric Transducer

In piezoelectric material, a voltage (signal) appears across the surface of the crystal, if the dimension of the crystal is changed by the application of a mechanical force. At the same time, vice versa is also true, i.e. converts from mechanical energy into electrical energy. These materials are usually ammonium dihydrogen phosphate, rochelle salt, quartz, dipotassium tartrate, potassium dihydrogen phosphate, lithium sulphate, etc. The piezoelectric effect is sensitive to direction. A compressive force produces voltage of one polarity whereas tensile force produces voltage of opposite polarity. The magnitude of the induced surface changes is proportional to applied force. The basic-equivalent circuit of piezoelectric transducer is as shown in Fig. 6.14.

The different modes of utilizing the piezoelectric transducer effect are: (i) volume expander, (ii) thickness shear and (iii) face shear. The various piezoelectric transducers are widely used for measurement of force, torque, pressure, acceleration, etc. Figure 6.15 shows an arrangement of piezoelectric transducers to measure the pressure under dynamic conditions and are used in lighter microphones, hydrophones and pressure indicators. In the given figure, the piezoelectric microphone consists of diaphragm, bimorph (piezoelectric crystal) and a spindle. The spindle connects the diaphragm and bimorph. The natural frequency of diaphragm, bimorph and other associated system must be kept higher than the highest responded input frequency. The prestressing of the pressure transducers is done by means of a thin-walled tube in order to enable the pressure fluctuations about a mean value to be measured. Here hollow spheres and hollow cylinders of ceramic materials (barium titanate) are used for measuring air blast pressure and underwater pressure.

Fig. 6.15 Piezoelectric
transducer to measure
pressure or convert sound to
electrical signal in mikes

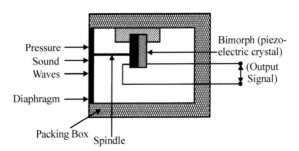

6.5.3 *Photoelectric Transducer*

It is an optoelectronic or optical transducer as shown in Fig. 6.16. Such transducer usually uses a phototube and light source separated by a small window, whose aperture is controlled by the force-summing device. The light source may be the ray of sun or bright LED depending upon the applications.

Fig. 6.16 Photoelectric transducer

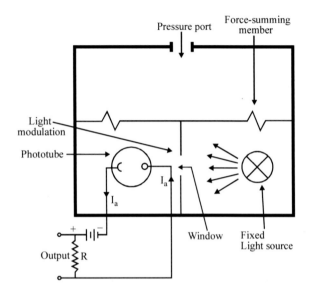

The quantity of incident ray (light) on the photosensitive cathode is varied in accordance to the externally applied force which reduces the windows size of incoming light, thereby changing the anode current I_a. Such devices measure and operate for both static and dynamic conditions. It has high efficiency but unable to respond for high-frequency light variation.

The other optoelectric transducers such as photovoltaic cell (solar cell) and photoconductive cell (LDR) have already been discussed in Chap. 4.

Questions

Fill up the blanks:

1. A transducer's function is to _____ energy.
2. In the absence of external power, the transducer that cannot work is called _____.
3. In the absence of external power, the transducer that can work is called _____.
4. Self-generation type transducers are called _____.

5. The property of a transducer to withstand loads is _____.
6. Displacement transducers convert the _____ force into displacement with the mechanical elements.
7. In wire-wound strain gauge, the change in resistance is due to _____.
8. A strain gauge is a passive transducer and is employed for converting _____.
9. _____ tends to elongate the wire and thereby increasing its length and decreasing its cross-sectional area.
10. Gauge factor/sensitivity of a strain gauge is _____.
11. Strain gauge is a _____ transducer.
12. LVDT is a _____ displacement transducer.
13. In a LVDT, the two secondary voltages are _____.
14. The thermistor has _____ and _____ temperature coefficient of resistance.
15. Hall effect can be used to find _____. It is an _____ transducer.
16. In Hall effect, the two input parameters _____ and _____ give rise to the third one _____.
17. The output of piezoelectric transducer is expressed as _____.
18. Piezoelectric crystals cannot be used for _____.
19. Thermocouples are generally used for temperature measurement up to _____.
20. Reference and sensing junctions are used in _____.
21. Capacitive transducer is _____ type transducer.
22. Analog transducers convert the input quantity into an analog output which is a _____ function of time.
23. Operating temperature range of thermistor is _____.
24. The electrical displacement transducer that works on the principle of variable inductance is called _____.
25. A thermocouple is a _____ transducer.

Short Questions:

1. What does transducer mean? How they are classified?
2. What are the active and passive transducers? Why are they called so? Give examples.
3. What are the basic requirements of transducers?
4. What is a thermocouple? Give examples.
5. What are the applications of transducer?
6. List out the various types of passive transducer.
7. What are the applications and features of thermistor?
8. Define accuracy and precision.
9. Mention the applications of strain gauge.
10. How do we get the two out of phase voltages from the two split windings of secondary of LVDT?

11. What is LVDT? What are the applications of LVDT?
12. What are the advantages and disadvantages of LVDT?
13. What are capacitor and inductive transducers?
14. What is piezoelectric transducer?
15. What are the suitable materials for piezoelectric transducer?
16. What is a Hall effect?
17. What is thermocouple?
18. What is photoelectric transducer?
19. What is potentiometric transducer?
20. What is active transducer?
21. Why does the conductivity of photoconductive material decreases with light?

Long Questions:

1. What is transducer? Give the basic block diagram of transducer.
2. When solar light falls on the solar cell, the electron–hole pair gets generated. Explain the mechanism by band theory.
3. With neat diagram explain the working principle and application of the following transducers:
   (a) Resistance strain gauge (bonded and unbonded),
   (b) Thermistor,
   (c) LVDT,
   (d) Hall effect transducer,
   (e) Inductive transducer and
   (f) Capacitive transducer.

4. Explain the working principle of the following active transducers with neat sketches:
   (a) Thermocouple,
   (b) Piezoelectric transducer
   (c) Photoconductive.

Chapter 7
Communication Systems

Contents

7.1 Communication

Simple communication is as old as human being's existence on earth. We communicate (*i.e.* exchange information) orally or in older times, pigeons were being sent for messages for longer distances or messengers were being sent for it. Today

© Springer Nature Singapore Pte Ltd. 2020   259
S. S. Srikant and P. K. Chaturvedi, *Basic Electronics Engineering*,
https://doi.org/10.1007/978-981-13-7414-2_7

these messages are being sent through electromagnetic (EM) waves as carriers of these low-frequency message signals. Therefore, voice waves and EM waves have a number of features we should know along with the frequency spectrum as given in the articles to follow.

7.1.1 Communication and Waves

Human being, animals, birds, etc., all communicate through voice which is mechanical wave, generated by some mechanical activity, e.g. movement of our vocal cord, tuning fork, reeds vibrating in ordinary harmonium, etc. The frequencies of voice and audibility of human being (as well as other living beings) differ a lot.

Normal voice frequency of male is 85–180 Hz and of female is 165–255 Hz, while singing frequency can be 77–482 Hz for male and 137–634 for female. Audibility of human being and animals differs a lot: human being (20 Hz–20 kHz), dogs (67 Hz–45 kHz), cats (55 Hz–79 kHz), Bats (10–200 kHz), Mice (1–70 kHz), Bird (1–4 kHz), Dolphin (75 Hz–150 kHz), elephant (17 Hz–10.5 kHz), mosquitoes (20–100 kHz), moth (20250 kHz), etc. Therefore, pest repellents having ultrasonic frequency from 20 to 40 kHz are kept in godowns.

The EM wave differs a lot with the mechanical wave. These can be compared in table given below:

Property	Mechanical wave	EM wave
1. Origin	It is due to some mechanical periodic disturbance	It is due to EM Wave
2. Generation	Produced by some mechanical vibration	Produced by vibration of charges, generating electric and magnetic fields which moves with velocity 'c'
3. Content	It has compression and rarefaction of a media, air, liquid or solids	Has varying electric and magnetic fields perpendicular to each other
4. Media	It needs a media to travel	Does not require any media to travel
5. Types	Three types: transverse, longitudinal and surface waves. In transverse the compression, rarefaction action of the media is to the direction of motion, while in longitudinal it is in the same direction. Surface wave are the mix of two e.g. Siesmic wave, Gravity wave etc	EM waves are of three types: TM, TE and TEM depending upon whether magnetic or electronic field is to the direction of motion or both
6. Velocity	Its velocity changes from media to media, on temperature, humidity pressure and its elastic properties. In dry normal (NTP) air, the sound velocity is 343 m/sec. This velocity increases with the frequency of sound	Velocity of EM wave is fixed in air/free space = c = 3×10^8 m/sec. In liquids or solids, velocity reduces

(continued)

(continued)

Property	Mechanical wave	EM wave
7. Energy	Mechanical wave energy depends on its amplitude, e.g. sound of explosive has very high amplitude due to very high compression–rarefaction wave, which even breaks the wall	It has energy = E = hν, *i.e.* Radio wave $E \leq 10^{-5}$ eV and microwave has ≈ 0.01 eV at 300 GHz. X-rays have = 10^3–10^5 eV
8. Frequency	(Human Audio) 20 Hz–20 kHz Ultrasonic: 20–50 kHz Birds: 1–4 kHz Mosquitoes 20–100 kHz	R.F. = 10 kHz to 300 GHz I.R. 10^{12} to 10^{14} Hz Light: 5×10^{14} to 10^{15} Hz UV ray: 10^{16} to 3×10^{17} Hz X-Ray: 3×10^{17} to 10^{20} Hz γ-Ray: 10^{18} to 10^{22} Hz

All mechanical waves can be converted into electronic signal (*i.e. ac* voltage) by sensors like mike. These electronic signals travel in wires with the velocity of light. These electronic signals can be converted into EM waves by transmitting through antenna, which travel through space with the velocity of light (3×10^8 m/s). These EM waves have electric (E) field and magnetic field (H) components perpendicular to each other. Reverse conversion of EM wave to mechanical wave in general is also possible for frequencies below 50 kHz or so. Therefore for frequencies <50 kHz:

Mech. waves $\leftrightarrow$ electronic signal $\leftrightarrow$ EM wave.

If the electronic signal has frequency more than 300 MHz then coaxial line or waveguides are convenient / effective media to travel.

7.1.2 Communication System and Frequency Spectrum of EM Waves

The communication services have been developed to send (or transfer) the information from a source to one or more destinations. Distribution of data, message or information from one location to another is the role of communication systems. The equipment that transmits the information is the transmitter and the equipment that receives the information is the receiver. The channel is the medium through which signal travels from the transmitter to the receiver. Few examples of communication services are telephony, telegraphy, facsimile (FAX), radio broadcast, TV transmission, computer communication, etc.

So, the basic function of a communication system is to communicate (or send or transfer) a message. The basic block diagram of a communication system is shown in Fig. 7.1.

The information or message to be transmitted by the information source is mostly non-electrical in nature. For example, audio signals in speech transmission and pictures in television transmission. This information in the original form is converted into a corresponding electrical signal known as the message signal by

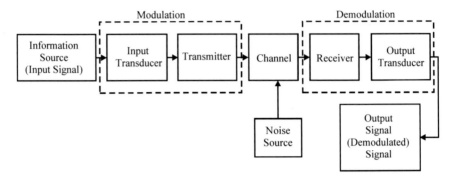

Fig. 7.1 Basic communication system

using an input transducer in the transmitter. This message signal cannot be directly transmitted due to various reasons, to be discussed in Sect. 7.4. Hence, this message signal is superimposed on a high-frequency carrier signal before transmission. This process is called as modulation. After modulation, the modulated carrier is amplified by using power amplifiers in the transmitter and fed to the transmitting antenna.

Channel is a medium through which the signal travels from the transmitter to the receiver. There are various types of channels, such as the atmosphere for radio broadcasting, wires for line telegraph telephony and optical fibres for optical communication. As the signal propagates through the channel, it gets attenuated by various mechanisms and affected by noise from the external source. Noise being an unwanted signal interferes with the reception, of wanted signal. In the design of communication system, careful attention should be paid to minimize the effect of noise on the reception of our wanted signals.

At the receiving end, a weak modulated carrier that is transmitted from the transmitter is received. This is first of all amplified to increase the power level and the process of demodulation is done to recover the original message signal from the modulated carrier. The recovered message signal is further amplified to drive the output device such as a loudspeaker or a TV receiver.

Table 7.1 gives the electromagnetic spectrum used for various communication services whereas Fig. 7.2 shows the comparative visualization of EM wavelengths and their frequencies.

7.2 Telecommunication Services

Telegraphy: This telecommunication service is used for transmission and reception of written texts. Teleprinters normally transmit signal at a speed of 50 bauds which occupies a bandwidth of 120 Hz. However, recent teleprinters can operate up to

Table 7.1 Electromagnetic spectrum for various communication services

Classification	Frequency	Wavelength	Uses
Very low frequencies (V.L.F.)	10–30 kHz	30–10 km	Long-distance point-to-point communications
Low frequencies (L.F.)	30–300 kHz	10–1 km	Long-distance point-to-point communication, marine, navigation. Power line carrier communication and broadcast
Medium frequencies (M.F.)	300–3000 kHz	1000–100 m	Power line carrier communication. Broadcast, marine communications, navigation and harbour telephone
High frequencies (H.F.)	3–30 MHz	100–10 m	Moderate and long-range communication of all types, broadcast
Very high frequencies (V.H.F.)	30–300 MHz	10–1 m	Short-distance communications. TV and FM broadcast, data communication. Mobile and navigation systems
Ultra-high frequencies (U.H.F.)	300–3000 MHz	1–0.1 m	Short-distance communications. TV broadcast, radar, mobile, navigation and microwave relay
Super-high frequencies (S.H.F.)	3–30 GHz	10–1 cm	Radar, microwave relay and navigation systems
Extremely high frequencies (E.H.F.)	30–300 GHz	1–0.1 cm (m m-wave)	Radar, satellite, mobile, microwave relay and navigations

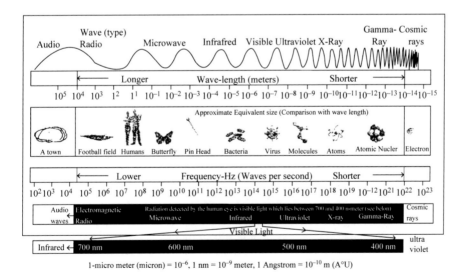

1-micro meter (micron) = 10^{-6}, 1 nm = 10^{-9} meter, 1 Angstrom = 10^{-10} m (A°U)

Fig. 7.2 Comparative visualization of the complete spectrum of EM wavelength and its frequencies

200 bauds. Teleprinters are interconnected through special exchanges called *telex* (telegraph exchange). These have now more or less become obsolete.

Telephony: This is a communication service for the transmission of speech signal between two points. The speech signal is converted into the corresponding electrical signal by using a microphone, and it is transmitted through a telephone line to a distant receiver where the original speech signal is reproduced by using a loudspeaker. Human voice produce harmonic signals in the band of 80 Hz–3.4 kHz. Hence, a bandwidth of 4 kHz is allocated for a telephone channel. A human ear is highly sensitive to sound between 3 kHz and 4 kHz. As the female voice contains more energy in this frequency range, they are preferred as telephone operators and announcers.

Facsimile (FAX): This is a telecommunication service for the transmission and reception of picture information like photographs, drawing, weather maps, etc. The picture or any document to be transmitted is mounted on a cylinder, and it is scanned by a photocell linked to the cylinder. The photocell produces electrical analog signal (as voltage variation depending upon the intensity of the light) and dark spots are produced on the document. The electrical signal thus produced is converted into frequency variations and transmitted through a telephone line. At the receiving end, the frequency variations are converted back into corresponding voltage variations that is given to a plotter for reconstructing the original picture or document. Thus, a photocopy of the original picture is obtained at a distance. Similar to a telephone signal, the FAX message also occupies a bandwidth of 4 kHz.

The other communication services like radio broadcast. TV transmission and computer communication are discussed in the later sections of this chapter.

7.2.1 Transmission Paths

Transmission of messages can be either through bound media such as pair of wires, coaxial cables, optical fibre cables, waveguides, etc., or through unbounded media such as free space or atmosphere.

Line Communication: Line communication refers to communication through pair of wires, coaxial cable and waveguides.

The *pair of wires* or *parallel-wire* is normally put/installed as overhead lines on poles and while the cables are normally buried under the ground. Buried cables have twisted pairs up to 4000 in numbers. The pairs are twisted to avoid crosstalk between subscribers. Such cables are used up to 500 kHz.

A coaxial line consists of a pair of concentric conductors with some dielectric filling the middle space, whereas the outer conductor is invariably grounded to act as an electrical shield. There may be a sheath around the outer conductor to prevent corrosion. A number of such coaxial lines are usually bunched inside a protective sleeve. Coaxial lines are employed for higher frequencies up to 18 GHz and a single coaxial line can be used to carry thousands of telephone channels.

A *waveguide* is used for signal transmission in the UHF range and SHF range, *i.e.* above 1 GHz. Here the signal gets propagated as an electromagnetic wave through a hollow pipe of rectangular or circular cross section. A waveguide acts a high-pass filter which does not pass the signal below its cut-off frequency. A waveguide has a bandwidth which is in excess of 20% of its operating frequency. As an example, a waveguide operating at 5 GHz offers a bandwidth of 1 GHz which can accommodate 2,50,000 telephone links.

An *optical fibre* is a waveguide used for transmitting signals in the optical frequency range from 10^{13} to 10^{15} Hz. Signal transmission through an optical fibre is based on total internal reflection.

Radio Communication: In radio communication, propagation of signals is through atmosphere or free space. Radio broadcasting, ground-based microwave communication and satellite communication are a few examples of this type, which are discussed in detail in the later sections of this chapter.

Radio waves of electromagnetic waves in the low-frequency (LF) region and medium frequency (MF) region are normally used for radio broadcasting. They are reflected by the different layers such as D, E, F of the ionosphere at a height of 50–400 km above the ground in the earth's atmosphere. Due to successive reflections from the ionosphere and the earth's surface (like that in waveguides), the radio waves travel a long distance and also increase the area of coverage of the broadcasting station.

Radio waves in the microwave frequency region above 1 GHz will penetrate the ionosphere, and hence, a satellite is required to reflect the signals towards the earth, which forms the basis for satellite communication.

7.3 Analog and Digital Signals

Signalling is the method through which information is transmitted across the medium. The information for the communication can be either of analog or digital form. The characteristics of an analog system are that it changes continuously (or instantaneously). The characteristics of a digital signal consists of discrete states like 'ON or OFF', 'Logic 0 or Logic 1', 'Yes or No', 'Open or Close' and so on.

Analog Signalling

An analog signal consists of electromagnetic waves which are changing continuously from maximum value to minimum and minimum to maximum and so on. Three characteristics are used to describe the analog signals, which has amplitude, frequency and phase. These three characteristics are shown in Fig. 7.3a.

The amplitude measures the capacity of the analog signal or the height of the wave, which is expressed in volts, current, watts and so on. The frequency is the times that it takes for a wave to complete one cycle and is measured in cycles per second or hertz. The phase sequence is the relative state of one wave when compared to another reference wave, which is measured in degrees.

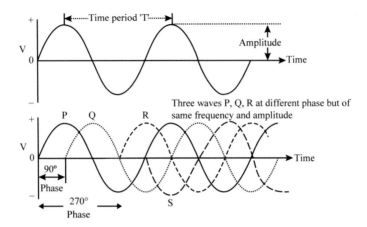

Fig. 7.3a Three characteristics of an analog signal—amplitude, frequency and phase

In Fig. 7.3a, the signal '*P*' is taken as reference, then signal '*Q*' is 90° phase shifted from '*P*' and signal '*R*' is 270° phase shifted from '*P*'. The signal '*S*' in 180° phase shifted from signal '*P*'.

Analog signals include the human voice, radio signal, the electrical signal (voltage or current), noise and so on. In an ordinary telephone loop, the signal consists of an electrical current which varies in amplitude in step with the variation of the intensity of sound that are impinging upon the microphone in the handset at the transmitting end. The variation of the electrical current amplitude is converted back to sound at the receiving end.

Digital Signalling: The digital data consists of two discrete states as shown in Fig. 7.3b with Logic 0 or Logic 1, ON or OFF, close or open, YES or No, and so on. Most of the computer networks use digital signalling. In digital, the state of change is practically instantaneous as shown in Fig. 7.3b. There are several ways to encode the digital data. With reference to Fig. 7.3b, the positive voltage is encoded as Logic '1' and the negative voltage is encoded as Logic '0'. This is referred as 'current state cooling'. Sometimes, Logic '1' can be encoded when transition taking place from low voltage to high voltage and Logic '0' can be encoded when transition taking place from high voltage to low voltage. This is called as 'state transition encoding.' Typically, the digital communication involves the transmission of 'ON/OFF' and 'Yes/No' system called bits.

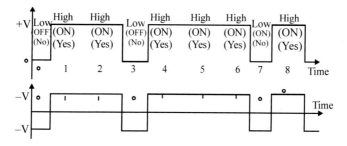

Fig. 7.3b Characteristics of digital signal

7.4 Basic Principle of Modulation

Modulation is the process of changing some parameter of a high-frequency carrier signal in accordance with the instantaneous variations of the message signal.

The carrier signal has a constant amplitude and frequency. The function of a carrier signal is to carry the message signal and hence the name.

The message or modulating signals are low-frequency audio or digital signals which contain the information to be transmitted. Generally, message signal ranges from 20 Hz to 20 kHz.

7.4.1 Need of Modulation

Modulation is an essential process in communication system to overcome the following difficulties is transmitting an unmodulated message signal.

(i) **Transmitting to Long Distance**: Low-frequency signal (e.g. audio) or digital signal cannot be send long distances in air requiring a high-frequency carrier signal to carry it

(ii) **Antenna Dimension**: When free space is used as communication media, messages are transmitted and received with the help of antennas. For effective transmission on reception, the dimension of the antenna should be of order quarter wavelength $(\lambda/4)$ of the signal that is transmitted. If an audio frequency signal is directly transmitted, the required dimension of the antenna will be quite large so it cannot be implemented in practice. For example, an audio frequency signal at 5 kHz required a vertical antenna of height $\lambda/4$.

where $\dfrac{\lambda}{4} = \dfrac{c}{f} = \dfrac{3 \times 10^8}{4 \times 5 \times 10^3} = 15,000$ m $= 15$ km (which is not practical)

Hence, modulation by a message signal of low-frequency carrier should be done so that it sits on the carrier of much higher frequency range and then transmitted, thereby reducing the required dimension of the transmitting and receiving antennas to a practical one.

(iii) **Interference**: In the audio frequency range of 20 Hz–20 kHz, the programmes of different stations will get mixed up and will be inseparable in the common communication channel. Hence to reduce the interference, modulation of message signals from different stations are done on different carrier frequencies, which transforms the modulated signal into different frequency slots.

(iv) **Channel Characteristics**: Different communication channels will sustain signals over different frequency ranges. A waveguide will sustain signals only in the microwave frequency range. An optical fibre will sustain signals only in the optical frequency range.

(v) **Ease of Bandwidth**: Since modulation translates the signal to higher frequencies, it becomes relatively easy to design and implement amplifiers and antenna systems.

(vi) **Adjustment of Bandwidth**: Signal-to-noise ratio in the receiver is a function of the bandwidth of the modulated signal. Bandwidth can be adjusted by the modulation process resulting in the improvement of signal-to-noise ratio.

(vii) **Shifting Signal Frequency to an Assigned Value**: Modulation process permits changing the signal frequency to a desired frequency band. This helps in effective utilization of the electromagnetic spectrum.

7.4.2 Types of Modulation

There are three types of modulation: (*i*) analog modulation, (*ii*) pulsed modulation and (*iii*) digital Modulation. They are further classified as below. For modulation, the types of signal/carriers are given in the Table 7.2a, b.

7.5 Various Modulation Method

The various modulation method, *i.e.* analog, pulsed and digital are discussed below. In general for both the modulation, the original information (or message) signal has to be carried by RF carrier signal in order to transmit information to the communication system. For this, the carrier is modulated by the message signal. This process of modulation is to change some parameter of a basic electromagnetic wave (RF) usually called the 'carrier wave'. In analog modulation method, analog carrier is used for getting modulated by the signal to be transmitted. In pulse digital modulation method, a digital carrier signal comprising a train of pulse, which is

Table 7.2a Types of modulation

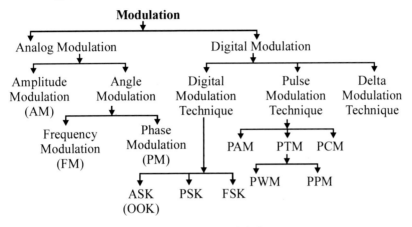

Note: 1. FM and PM are also called angle modulation.

2. PWM and PPM are also called pulse time modulation (PTM).

3. ASK is also called ON/OFF Keying as the modulated carrier looks like as if R.F. signal is put ON or OFF (Fig. 7.5)

Analog Mod: AM, FM, PM

Pulse Mod: PAM → Pulse Amplitude Modulation

                        PWM → Pulse Width Modulation

                        PPM → Pulse Position Modulation

                        PCM → Pulse Code Modulation

Digital Mod: ASK → Amplitude Shift Keying

                        FSK → Frequency Shift Keying

                        PSK → Phase Shift Keying

Table 7.2b Types of message signal and carrier in modulation

Carrier	Analog modulation (AM, FM, PM)	Pulsed modulation (PAM, PWM, PPM, PCM)	Digital modulation (ASK, FSK, PSK)
Message signal (low-frequency signal or data)	Analog	Analog	Digital
Carrier (high frequency)	Analog RF	Pulse train of RF shoots	Analog RF

used for getting modulation by the signal to be transmitted. In digital modulation method, digital data signal to be transmitted is used to modulate a sine wave carrier (see Table 7.2b).

7.5.1 Analog Modulation

As discussed earlier for analog modulation, the analog carrier (of high frequency) is used for getting modulated, *i.e.* for carrying message signal, which is also analog but of much lower frequency. There are three types of analog modulation (Figure 7.4A). They are (1) amplitude modulation, (2) frequency modulation and (3) phase modulation. In fact these two modulation, e.g. FM and PM are also called angle modulations.

7.5.1.1 Amplitude Modulation (AM)

In amplitude modulation, the amplitude of the high-frequency carrier signal (Fig. 7.4A(*a*)) is varied in accordance with the instantaneous value of the modulating signal as shown in Fig. 7.4A(*b*), but its frequency remains the same. The envelope of the modulated carrier is an exact replica of the audio frequency signal wave (Fig. 7.4A(*c*))

Fig. 7.4A Analog modulations: AM, PM and FM waves. **a** Carrier wave, **b** sinusoidal modulating wave, **c** amplitude-modulated wave, **d** phase-modulated wave and **e** frequency-modulated wave

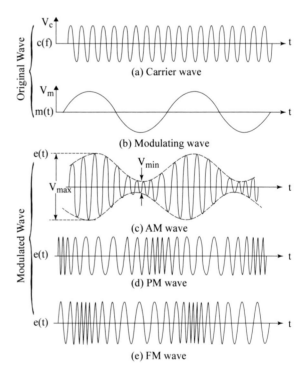

Let the carrier signal be represented by

$$c(t) = V_c \cos \omega_c t$$

where V_c is the constant amplitude of the carrier signal and the modulating signal is represented by
    i.e.

$$m(t) = V_m \cos \omega_m t$$

where V_m is the amplitude of the modulating signal.
    The instantaneous voltage of the resulting amplitude-modulated carrier (ω_c) wave is represented by

$$e(t) = A \cos \omega_c t$$

The amplitude A of the amplitude-modulated (AM) wave will be:

$$A = V_c + m(t) = V_c + V_m \cos \omega_m t$$

Hence

$$e(t) = (V_c + V_m \cos \omega_m t) \cos \omega_c t$$
$$= V_c \left[1 + \frac{V_m}{V_c} \cos \omega_m t \right] \cos \omega_c t \qquad (7.1)$$
$$= V_c [1 + m_a \cos \omega_m t] \cos \omega_c t$$

where $m_a = \frac{V_m}{V_c}$ modulation index (depth of modulation using this we can know also the percentage of modulation) of the amplitude modulation. For $m_a = 1$, it is full (*i.e.* 100%) modulation. From Fig. 7.4A(c) with V_{min} and V_{max} as the minimum and maximum amplitude of the modulated signal, we know that

$$V_m = \frac{V_{max} - V_{min}}{2}$$
$$V_c = V_{max} - V_m = V_{max} - \frac{V_{max} - V_{min}}{2}$$
$$= \frac{V_{max} + V_{min}}{2}$$

Therefore

$$m_a = \frac{V_m}{V_c} = \frac{V_{max} - V_{min}}{V_{max} + V_{min}}$$

So, the value of m_a should lie between 0 and 1, and $e(t)$ of equation no. 7.1 becomes

$$e(t) = V_c \cos \omega_c t + m_a V_c \cos \omega_c t \cos \omega_m t$$

$$e(t) = V_c \cos \omega_c t + \frac{mV_c}{2} [\cos(\omega_c + \omega_m)t + \cos(\omega_c - \omega_m)t] \qquad (7.2)$$

Hence, the AM wave consists of three frequency components (Fig. 7.4B(a)):

(i) f_c, the carrier frequency components.
(ii) $f_c + f_m$, the sum frequency component called as upper sideband (USB) f_{USB} frequency component.
(iii) $f_c - f_m$, the difference frequency component called as lower sideband (LSB) f_{LSB} frequency component.

Power Relation in the AM Wave: The total power (P_t) in the modulated wave, shown in Fig. 7.4A(c) is equal to the sum of the unmodulated carrier power and two sideband powers, which are equal:

$$P_t = P_c + P_{LSB} + P_{USB}$$

$$P_t = \frac{V_{c-RMS}^2}{R} + \frac{V_{LSB}^2}{R} + \frac{V_{USB}^2}{R}$$

where all these voltages are RMS values of carrier, LSB and USB, while R is the antenna resistance (Fig. 7.4B), with $V_{LSB} = V_{USB} = m_a V_c/2$; $V_{c-RMS} = V_c/\sqrt{2}$; $V_{SB-RMS} = (m_a V_c/2)/\sqrt{2}$

The unmodulated power is, $P_c = \dfrac{V_{c-RMS}^2}{R} = \dfrac{\left(V_c/\sqrt{2}\right)^2}{R} = \dfrac{V_c^2}{2R}$

Similarly, $P_{LSB} = P_{USB} = \dfrac{V_{SB-RMS}^2}{R} = \dfrac{(m_a V_c/2)^2}{2R} = \dfrac{m_a^2 V_c^2}{8R} = \dfrac{m_a^2}{4} \dfrac{V_c^2}{2R}$ $\qquad (7.3)$

Therefore $P_t = \dfrac{V_c^2}{2R} + \dfrac{m_a^2}{4\cdot} \dfrac{V_c^2}{2R} + \dfrac{m_a^2}{4\cdot} \dfrac{V_c^2}{2R} = P_c + \dfrac{m_a^2}{4} P_c + \dfrac{m_a^2}{4} P_c = P_c(1 + \dfrac{m_a^2}{2})$

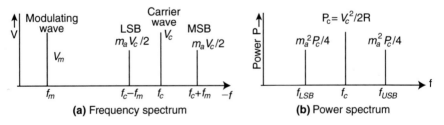

(a) Frequency spectrum   (b) Power spectrum

Fig. 7.4B Amplitude modulation. **a** carrier and modulating wave together creates LSB (f_{LSB}) and USB (f_{USB}) signal voltages in frequency spectrum, **b** power in frequency spectrum

Therefore

$$\frac{P_t}{P_c} = 1 + \frac{m_a^2}{2} \qquad (7.4)$$

The maximum power in the AM wave is $P_t = 1.5\,P_c$, when $m_a = 1$.

Current Calculations: Let the RMS values of the unmodulated current and the total modulated current of an AM transmitter be I_c and I_t, respectively, then

$$\frac{P_t}{P_c} = \frac{I_t^2 R}{I_c^2 R} = \left(\frac{I_t}{I_c}\right)^2 = 1 + \frac{m_a^2}{2}$$

Therefore

$$\frac{I_t}{I_c} = \sqrt{1 + \frac{m_a^2}{2}}$$

Hence

$$I_t = I_c \sqrt{1 + \frac{m_a^2}{2}} \qquad (7.5)$$

Modulation by Several Sine Waves: Let V_1, V_2, V_3, etc., be the simultaneous modulating voltages. Then the total effective modulating voltage V_t is equal to the square root of the sum of the squares of the individual voltages, *i.e.*

$$V_t = \sqrt{V_1^2 + V_2^2 + V_3^2 + \dots}$$

Dividing throughout by V_c, we obtain

$$\frac{V_t}{V_c} = \frac{\sqrt{V_1^2 + V_2^2 + V_3^2 + \dots}}{V_c}$$

$$\frac{V_t}{V_c} = \sqrt{\frac{V_1^2}{V_c^2} + \frac{V_2^2}{V_c^2} + \frac{V_3^2}{V_c^2} + \dots}$$

Hence, the total modulation index is

$$m_{at} = \sqrt{m_{a1}^2 + m_{a2}^2 + m_{a3}^2 + \dots} \qquad (7.6)$$

7.5.1.2 Angle Modulation

In angle modulation, the instantaneous angle of the carrier signal is varied in accordance with the instantaneous value of the modulating signal and its amplitude is kept constant. Two forms of angle modulation are (i) frequency modulation and (ii) phase modulation.

7.5.1.3 (a) Frequency Modulation (FM)

In frequency modulation, the instantaneous frequency of the carrier is varied linearly with the variations of the message signal while the amplitude of the modulated carrier remains constant as shown in Fig. 7.4e. The instantaneous frequency of the frequency-modulated carrier can be written as

$$f_i(t) = f_c + k_f m(t)$$

where $f_i(t)$ varies linearly with the baseband signal, k_f is the frequency sensitivity constant (Hz/V) and f_c is the unmodulated carrier frequency.

Therefore

$$\omega_i(t) = 2\pi f_c + 2\pi k_f m(t)$$

Then,

$$\theta_i(t) = \int \omega_i(t)\, dt$$

$$= 2\pi f_c \int dt + 2\pi k_f \int m(t)\, dt$$

$$\theta_i(t) = 2\pi f_c t + 2\pi k_f \int m(t)\, dt$$

So the frequency-modulated wave is represented by

$$e(t) = V_c \cos\left[2\pi f_c t + 2\pi k_f \int m(t)\, dt\right] \tag{7.7}$$

When the modulating signal is $m(t) = V_m \cos \omega_m t$, then:

$$e(t) = V_c \cos\left[2\pi f_c t + \frac{2\pi k_f V_m \sin \omega_m t}{\omega_m}\right]$$

$$= V_c \cos\left[2\pi f_c t + \frac{k_f V_m \sin \omega_m t}{f_m}\right]$$

$$e(t) = V_c \cos\left[2\pi f_c t + \frac{\delta \sin \omega_m t}{f_m}\right] \tag{7.8}$$

where $\delta(=k_f V_m)$ is defined as the peak (maximum) frequency deviation.

$$e(t) = V_c \cos\left[\omega_c t + m_f \sin \omega_m t\right]$$

Hence, the modulation index for FM, m_f is defined as

$$m_j = \frac{\text{maximum frequency deviation}}{\text{modulating frequency}} = \frac{\delta}{f_m} \qquad (7.9)$$

The maximum value of frequency deviation (δ) is fixed at 75 kHz for commercial FM broadcasting. For normal band, the value of modulation index (m_f) is less than 1. For wideband FM, m_f is greater than 1.

Unlike AM, where there are only the carrier and the two sideband components, the FM wave consists of a single carrier frequency component and an infinite number of side frequency components. Hence, the bandwidth required for transmission of FM signal is larger than for the AM signal. However, as the amplitude of the carrier is kept constant, FM signal is less affected by noise than the AM signal. Table 7.3 shows a detailed comparison between AM and FM signal with negative (−ve/⊖) and positive (+ve/⊕) properties.

7.5.1.4   (b) Phase Modulation (PM)

In phase modulation, the phase of the carrier is varied in accordance with the instantaneous value of the modulating signal, whereas the amplitude of the modulated carrier is kept constant, as shown in Fig. 7.4d. The phase angle cannot change without affecting the frequency. So phase modulation and frequency modulation are interrelated with each other.

In phase modulation, the instantaneous angle of the phase-modulated carrier $\theta_i(t)$ is

$$\theta_i(t) = \omega_c t + k_p m(t)$$

where $\omega_c t$ is the angle of the unmodulated carrier and k_p is the phase sensitivity. $\theta_i(t)$ varies linearly with the base band signal. Hence, the phase-modulated wave is represented as

$$\begin{aligned} e(t) &= V_c \cos[\theta_i(t)] \\ &= V_c \cos\left[\omega_c t + k_p m(t)\right] \end{aligned}$$

Table 7.3 Comparison of AM/FM (−ve and +ve properties)

Parameters	AM	FM
(1) Frequency range	935–1705 kHz (or 1200 bits/s)	88–108 MHz (or 1200–2400 bits/s)
(2) Bandwidth for a signal to be transmitted	⊕ Modulating signal has bandwidth of 15 kHz, AM signal require 30 kHz bandwidth	⊖ It is BW $= 2(f_m + \Delta f_c)$ That is, twice the sum of modulation and carrier frequency deviation $\therefore$ Bandwidth for $\Delta f_m = 15$ kHz $\Delta f_c = 75$ kHz $\therefore$ Bandwidth $= 2(15 + 75)$ $= 180$ kHz
(3) Sound Quality	⊖ Poor quality	⊕ Better quality
(4) Noise	⊖ More susceptible to noise	⊕ Less susceptible to noise
(5) Circuit	⊕ Transmitter and receiver circuits are simple	⊖ Transmitter and receiver circuits are a bit complicated
(6) Stations	⊕ Every 10 kHz, one station is possible. So, more stations/channels are possible	⊖ Every 200 kHz, we can have one station. So less channels are possible
(7) Distance	⊕ Ionospheric reflection is used, and therefore can go to 140–700 km. Also high frequency will travel longer	⊖ FM transmitter works on line of sight and therefore can transmit up to 80–100 km only
(8) Fading	⊖ AM is transmitted through the air and ionosphere and therefore less stable due to variation of ionosphere with time	⊕ Due to line-of-sight transmission, fading is not there
(9) Power	⊖ Higher power of transmitter is required for the same distance due to (a) low frequency, (b) ionosphere and (c) attenuation	⊕ As compared to AM, much less power of transmitter is required for the same distance
(10) Modulating index	Modulating index is proportional to the modulating voltage only	Modulating index is proportional to the modulating voltage and is inversely proportional to the modulating frequency

Substituting $m(t) = V_m \cos \omega_m t$, we get:

$$e(t) = V_c \cos\left[\omega_c t + k_p V_m \cos \omega_m t\right] \\ = V_c \cos\left[\omega_c t + m_p \cos \omega_m t\right] \tag{7.10}$$

where $m_p = k_p V_m$ is the modulation index for phase modulation, which is the maximum phase deviation of the phase-modulated carrier.

Table 7.3 shows the comparison of FM and AM.

7.5.2 Digital Modulation

Digital signals (which consists of 'Logic 0 or Logic 1') for transmitting over long distances, digital modulation techniques like amplitude-shift keying (ASK), frequency-shift keying (FSK) and phase-shift keying (PSK) are employed as shown in Fig. 7.5 by varying the amplitude, frequency or phase of a sinusoidal carrier wave, respectively.

7.5.2.1 Amplitude-Shift Keying (ASK) or ON–OFF Keying (OOK)

Such modulation is achieved by varying the amplitude of the signal. A carrier wave is switched ON (corresponding to Logic '1') and OFF (corresponding to Logic '0') by the binary signals. In other words, the carrier wave is transmitted during Logic '1' and the carrier wave is suppressed during Logic '0' as shown in Fig. 7.5. Sometimes stronger amplitude of the signal referred as Logic '1' and weaker (ideally zero) amplitude of the signal referred as Logic '0' as shown in Fig. 7.5d, frequency (ω) remaining the same (Fig. 7.5a, d).

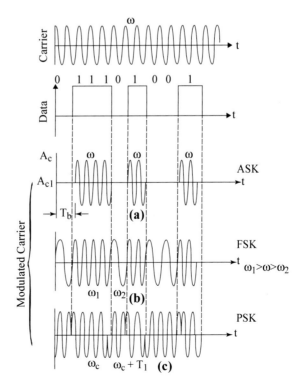

Fig. 7.5 Three basic forms of digital modulation: **a** amplitude-shift keying, **b** phase-shift keying and **c** frequency-shift keying

Fig. 7.5d ASK showing Logic '1' as stronger amplitude and Logic '0' as weaker amplitude, with frequency unchanged

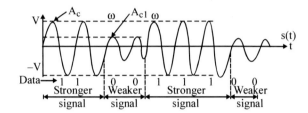

Mathematically for ASK,

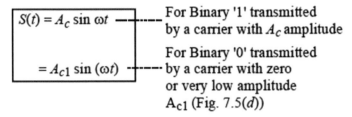

$$S(t) = A_c \sin \omega t$$
For Binary '1' transmitted by a carrier with A_c amplitude

$$= A_{c1} \sin (\omega t)$$
For Binary '0' transmitted by a carrier with zero or very low amplitude A_{c1} (Fig. 7.5(d))

7.5.2.2 Frequency-Shift Keying (FSK)

In FSK, two different carrier frequencies (ω_1 and ω_2) are used and they are switched ON and OFF by the binary signals (Fig. 7.5b) Binary '1' is transmitted as a carrier of particular frequency (f_1 or angular frequency ω_1) and binary '0' is transmitted is a carrier of different frequency (f_2 or angular frequency ω_2). Mathematically, it is represented as

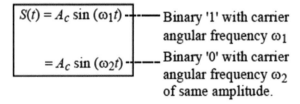

$$S(t) = A_c \sin (\omega_1 t)$$
Binary '1' with carrier angular frequency ω_1

$$= A_c \sin (\omega_2 t)$$
Binary '0' with carrier angular frequency ω_2 of same amplitude.

7.5.2.3 Phase-Shift Keying (PSK)

In PSK, the binary signals are used to switch the phase of a carrier wave between two values 0° and 180°. The positive phase is referred as Logic '1' in which it is transmitted as a carrier of particular frequency (*i.e.* phase angle $\theta_1 = \omega_c t$). The reverse phase is referred as Logic '0' in which it is transmitted as a carrier of same frequency as well as same amplitude but with a 180° phase shift (*i.e.* phase angle $\theta_2 = \omega_c t + \pi$). It is shown in Fig. 7.5c.

Mathematically, it is represented by

$$S(t) = A_c \sin \theta_1 = A_c \sin \omega_c t \qquad \text{------- Binary '1'}$$

$$= A_c \sin \theta_2 = A_c \sin (\omega_c t + \pi) = -A_c \sin \omega_c t \qquad \text{------ Binary '0'}$$

7.5.3   Pulse Modulation

In pulse modulation, the digital carrier signal is modulated by an analog modulating signal. Pulse modulation techniques are very important technique, when it is desired to transmit a large number of signal simultaneously through a single channel in an efficient manner. This modulation techniques yield better signal-to-noise ratio (S/N) at the receiving end, and hence, they are highly immune to noise.

Here, a train of rectangular pulses is considered to be a carrier signal as shown in Fig. 7.6b. In this modulation technique, the continuous waveform of the message signal is sampled at regular intervals. Information regarding the message signal is transmitted only at the sampling times. So, for proper recovery of the message signal at the receiving end, the sampling rate should be greater than a specified value which is given by sampling theorem.

7.5.3.1   Sampling Theorem

The sampling theorem states that if the sampling frequency (f_s) of any pulse modulation system exceeds twice the maximum signal modulating frequency (f_m), then the original signal reaches the receiving end with minimal distortion. Hence, the condition, $f_s \geq 2f_m$ should be satisfied in any pulse modulation system. For example, for a standard telephone channel which consists of audio frequency signals in the range of 300–3400 Hz, a sampling rate of 8000 samples per second (here $2 \times 3400 = 6800$ Hz) is a worldwide standard.

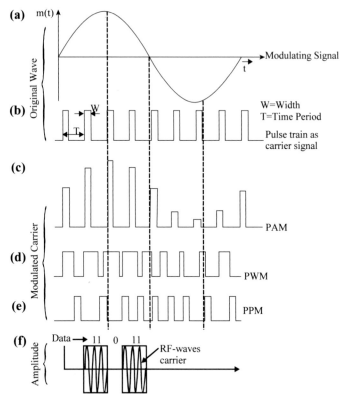

Fig. 7.6 Pulse analog modulation techniques: **a** modulating wave, **b** pulse carrier, **c** PAM wave, **d** PWM wave, **e** PPM wave and **f** two sample pulses as RF shoots for actual transmission of 11011 through air as media

7.5.3.2 A Transmitting Pulse-Modulated Wave

Since a pulse-modulated wave when analysed by Fourier analysis, it will show to have *dc* and low-frequency components, and therefore, direct transmission is not possible. Therefore, a RF carrier of high frequency will be required. Therefore, each rectangular pulse will have to be shoots of RF waves, then only it can be transmitted through air [see Fig. 7.6f].

7.5.3.3 Pulse Amplitude Modulation (PAM)

PAM is the simplest type of pulse modulation as shown in Fig. 7.6c. In PAM, the amplitude, of each pulse of the unmodulated pulse train is varied in accordance with the sample value of modulating signal.

7.5.3.4 Pulse Time Modulation (PTM)

In PTM, the amplitude of the pulses of the carrier pulse train is kept constant but their timing characteristics are varied and made proportional to the sampled signal amplitude at that instant. The variable timing characteristics may be pulse width or position, leading to pulse width modulation or pulse position modulation.

(a) **Pulse Width Modulation (PWM)**
PWM is also called as pulse duration modulation (PDM) or pulse length modulation (PLM). It is shown in Fig. 7.6d in which the amplitude and starting time of each pulse is fixed, but the width of each pulse is made proportional to the amplitude of signal at that instant. The pulses of PWM are of varying width and therefore of varying power content. Even if synchronization between transmitter and receiver fails, PWM still works whereas PPM does not.

(b) **Pulse Position Modulation (PPM)**
In pulse position modulation (PPM) as shown in Fig. 7.6e, the amplitude and width of the pulses are kept constant but the position of each pulse is varied in accordance with the instantaneous sample value of the modulating wave.

7.5.3.5 Pulse Code Modulation (PCM)

PCM is a pulse digital modulation technique in which the analog message signal is converted into a digital signal before transmission. The essential operations in the transmission of a PCM systems are sampling, quantizing and encoding as shown in Fig. 7.7a.

Sampling: The incoming message signal is sampled with a train of narrow rectangular pulses. The rate of sampling should be greater than twice the highest frequency component of the message signal in accordance with the sampling theorem.

Quantizing: Quantizing is a process in which the sample value obtained at any sampling time is rounded off to the nearest standard (or quantum) level as shown in Fig. 7.5b. For example, when the sample value is 6.8 V, it can be rounded off to 7 V.

(a)

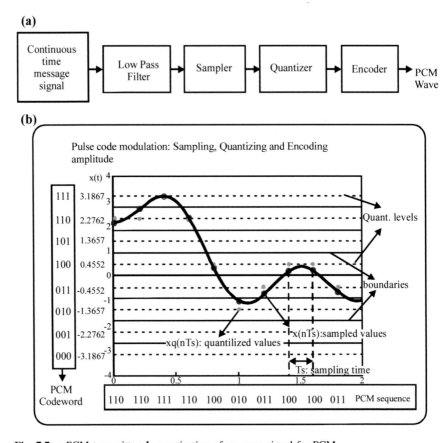

Fig. 7.7 **a** PCM transmitter; **b** quantization of message signal for PCM

Encoding: This process translates the quantized values into a corresponding binary word by choosing an appropriate coding format. Thus, the quantized value 7 can be converted as the binary word 0111 by using the 8421 code. If the number of quantum levels is 16 (2^4), four binary digits are required to represent a sample. This binary sequence is transmitted to the receiving end, by using a RF carrier by any of the three digital modulation method.

7.6 Transmitter and Receiver

7.6.1 AM Transmitter

Figure 7.8 shows a typical block diagram of an AM transmitter, where either low- or high-level modulation is employed. As shown in the block diagram, the audio

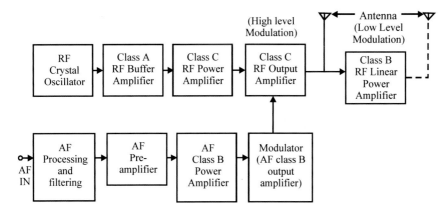

Fig. 7.8 AM transmitter block diagram

signal is filtered in order to occupy the correct bandwidth and amplified by two or three stages of audio frequency (AF) amplifiers. The radio frequency (RF) carrier wave generated by the three stages of amplifiers, to generate the required output power.

For high-level modulation, the amplified audio frequency signal is applied in series with the collector circuit of class C of the RF output amplifier transistor. The generated AM wave is directly given to the transmitting antenna. The purpose of using class C amplifier as that it has maximum efficiency.

For low-level modulation, the amplified AF signal is applied in series with the base circuit of Class C of the RF transistor amplifier, and the generated AM wave is amplified by using a Class B RF linear power amplifier and fed to the transmitting antenna. Broadcast transmitters invariably use high-level modulation because of the larger amount of power requirements.

7.6.2   FM Transmitter

The block diagram of a directly modulated PM transmitter is shown in Fig. 7.9. The transmitter employs a reactance modulator (varactor diode) to produce a frequency deviation in proportion to the signal amplitude at the output of the LC oscillator. The reactance modulator varies the total reactance of the resonant circuit of the LC oscillator thereby varying its output frequency. The resulting FM wave is passed through a number of frequency multiplier stages. These stages raise the centre frequency as well as the frequency deviation by the same factor. The frequency-modulated wave is then amplified to the required power level by class C power amplifier stages and transmitted.

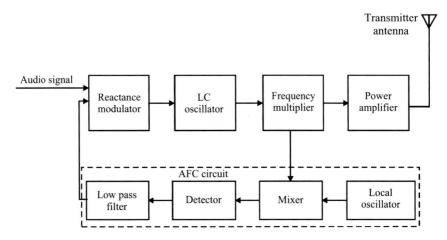

Fig. 7.9 Block diagram of FM transmitter

The automatic frequency control (AFC) circuit maintains the centre frequency of the transmitter output at a fixed value if there is any drift in it due to changes in circuit parameter due to change in temperature, etc. Signal from the frequency multiplier is mixed with the local crystal oscillator output and the difference frequency is fed to a discriminator (FM demodulator). The discriminator gives a *dc* output according to the frequency shift with respect to the centre frequency and goes to the reactance modulators (varactor diode) to bring the centre frequency of LC oscillator output back to its original value. Frequency-modulated radio (FM radio) uses the ultra-short wave band from 88 to 108 MHz. The channel spacing is 300 kHz. The FM reception is better than that of AM because FM is immune to noise and atmospheric disturbances.

7.6.3 Concept of Superheterodyning and Intermediate Frequency

There are basically two types of radio receivers that have commercial significance: (*i*) tuned radio frequency (TRF) receiver and (*ii*) superheterodyne receiver.

TRF Receiver

A TRF receiver, as shown in Fig. 7.10, perform the functions of interception, selection, RF amplification, detection, audio amplification and works satisfactorily at medium-wave frequencies. At higher radio frequencies, the performance of TRF

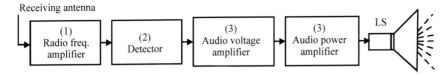

Fig. 7.10 TRF receiver

becomes poor. The performance of receiver is improved by a technique know as heterodyning. A receiver based on this technique is called superheterodyning receiver. The performance of a receiver is judged from its various features such as selectivity, sensitivity and fidelity.

(i) **Selectivity**: It is receiver's ability to distinguish between two adjacent carrier frequencies. By this feature, it is decided how perfectly the receiver is able to select the desired carrier frequency reject the others. Selectivity depends upon the sharpness of the resonance curve of tuned circuits involved in the receiver. The sharper the resonance curve, the better is the selectivity. Better selectivity means that the receiver has greater capability to reject undesired signals. The sharpness of the resonance curve depends on Q of the resonant circuit. The higher the Q, the more is the electivity. In Fig. 7.11, the curve 1 is sharper than the curve 2, and hence has better selectivity.

(ii) **Sensitivity**: The ability of a receiver to detect the weakest possible signal is called sensitivity. A receiver with a good sensitivity will provide more output for a similar input signal as against a receiver with poor sensitivity. The sensitivity of a receiver is decided by the gain of its amplifying stages.

(iii) **Fidelity**: The ability of a receiver to reproduce faithfully all frequency components present in the baseband signal is called fidelity, *i.e.* wideband. If any component is missed, or attenuated considerably, fidelity suffers and the reproduced signal is distorted. This feature is mainly decided by the bandwidth of audio amplifier which amplifies the baseband signal.

Superheterodyne Receiver

This receiver is popularly used in commercial application. The heterodyning gives a far better performance than the TRF receiver. The block diagram of the basic superheterodyne receiver in shown in Fig. 7.12a and full diagram at Fig. 7.12b.

Fig. 7.11 Resonance curve of a tuning circuit

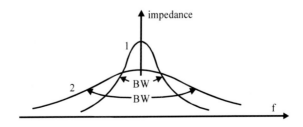

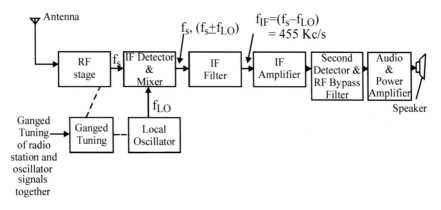

Fig. 7.12(a) Block diagram of superheterodyne receiver

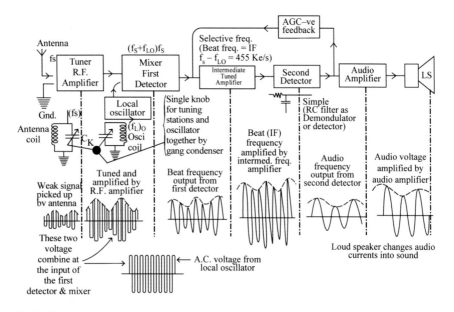

Fig. 7.12(b) Superheterodyne receiver used in radio sets: difference of antenna signal frequency and frequency of oscillator has to remain constant = IF = 455 kHz and for this tuning circuit of both has to change the value of capacitor (C) together in a gang condenser by a single timing knob K. AGC is a −ve feedback for signal level stabilization of audio signal output

The modulated signal transmitted from the broadcast station is picked up by the antenna and fed to the RF stage where signal is amplified. The signal voltage is combined with the output from a local oscillator in the mixer and station tuning done such that the difference frequency is a signal of lower fixed frequency called intermediate frequency (IF). The process is referred as superheterodyning (see Fig. 7.12a).

The **need for a superheterodyning** arose because of the fact that no single amplifier can be so wideband amplifier to cover large frequency range of radio stations, e.g. medium-wave (MW) (540–1640 kHz or Kc/s), short-wave (SW) (3–18 MHz or Mc/s) and frequency-modulated (FM) (88–108 MHz or Mc/s) signals. The **working of a superheterodyne receiver** can be clearly understood from elaborated way from Fig. 7.12b. As per the principle of superheterodyne receiver shown in Fig. 7.12b, the signal of antenna (f_s) and the signal from local oscillator (f_{LO}) is mixed in the 1st detector to give its output as $f_s, (f_s \pm f_{LO})$. This goes to the tuned intermediate frequency (IF) amplifier, which tunes and amplifies the signal containing the frequency $(f_s - f_{LO}) = 455$ kHz only. This is possible by a gang condenser having two parts each working as LC tuners each for antenna signal/coil and for oscillator (generator) signal/coil. **Therefore, radio station frequency tuning and oscillator frequency tuning happen together, so that the difference of these two frequencies (IF) remains 455 kHz always. This IF signal can easily be amplified with very high gain by a frequency tuned IF amplifier.**

7.7 Satellite Communication

A satellite is a radio repeater, also called transponder, placed in the sky. A satellite system as shown in Fig. 7.13a consists of a transponder and a minimum of two earth stations, one for transmission and other for reception. The transponder receives the signal from the transmitting earth station, frequency converts, amplifies and retransmits the signal towards the receiving earth station. The satellite are generally classified as passive and active types. A passive satellite simply reflects a signal back to earth and there is no gain devices on board to amplify the signal. On the other hand, an active satellite receives, amplifies and retransmits the signal back towards earth (Fig. 7.13b).

Once launched, a satellite remains in orbit as the centrifugal force caused by its rotation around the earth is counterbalanced by the earth's gravitational pull. Closer to earth the satellite rotates, the greater the gravitational pull, and greater the velocity required to keep it from being pulled to earth.

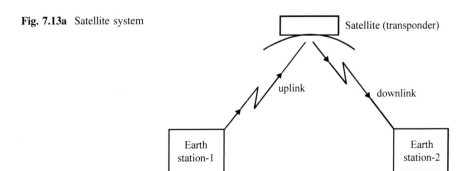

Fig. 7.13a Satellite system

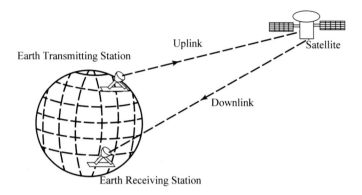

Fig. 7.13b Satellite communication/transmission system

(a) **Low altitude satellites (150–500 km in height)**: The satellites that orbit close to the earth travelling at approximately 28,500 km/hr are called low altitude satellites. At this speed, it takes approximately 90 min to rotate around the entire earth. Therefore, the satellite is in the line of sight (LOS) of a particular earth station for only 15 min or less per orbit.

(b) **Medium altitude satellites (9,500–19,000 km in height)**: These have a rotation period of 5–12 h and remain in LOS of a particular earth station for 2–4 h per orbit.

(c) **High altitude, geosynchronous satellite (at 35,786 km height)**: A satellite which has an orbital velocity equal to that of the earth's and are at a height of 35,786 km are geosynchronous. A geosynchronous satellite that lies on the earth's equatorial plane is called as the *geostationary satellite*. A geostationary satellite remains in a fixed position with respect to a given earth station and has 24 h availability time. Three geostationary satellites spaced 120° apart can cover the whole world.

Based on the area of the coverage, the orbits are classified as (*i*) *polar orbit*, (*ii*) *inclined orbit* and (*iii*) *equatorial orbit*. When the satellite remains in an orbit above the equator, it is called an equatorial orbit. When the satellite rotates in an orbit that takes it over the north and south poles, it is called polar orbit. Any other orbital path is called an inclined orbit. 100% of the earth's surface can be covered with a single satellite in a polar orbit. This is because the satellite is rotating around the earth in a longitudinal orbit, while the earth is rotating on a latitudinal axis; therefore, every location on earth lies within the radiation pattern of the satellite twice each day.

A satellite system consists of three basic sections; (*a*) the uplink (transmitting earth station) (*b*) the satellite transponder and (*c*) the downlink (receiving earth station). Typical frequencies for telecommunication services in a satellite system for uplink/downlink are 6/4 GHz or 14/12 GHz, where 6 and 14 GHz represent uplink frequencies and 4 and 12 GHz, the downlink frequencies.

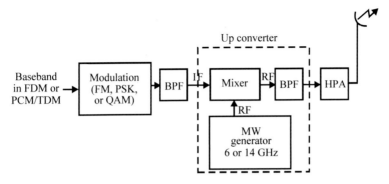

Fig. 7.14 Block diagram of earth station transmitter (or uplink converter section)

(a) **Uplink Converter**: The primary component of the uplink section of a satellite system is the earth station transmitter. The Fig. 7.14 is the block diagram representation of the earth station transmitter, which normally consists of an IF modulator, an IF to RF microwave upconverters, a high power amplifier (HPA) and a band-pass filter (BPF). The IF modulator converts the input baseband signals to either an FM or a PSK-modulated intermediate frequency. The upconverter translates the IF to an appropriate RF carrier frequency. The HPA provides adequate output power to propagate the signal to the satellite transponder. Normally used HPAs are klystrons and travelling wave tube amplifier.

(b) **Satellite Transponder**: The satellite transponder shown in Fig. 7.15 consists of a band-pass filter (BPF), an input low noise amplifier (LNA), a frequency translator and a high power amplifier (HPA). The transponder is an RF to RF repeater, of receiving RF to slightly lower RF transmission.

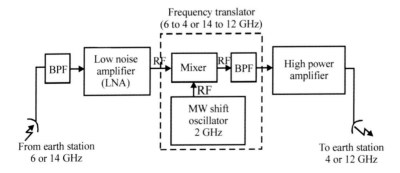

Fig. 7.15 Block diagram of the satellite transponder

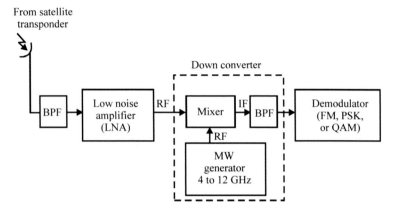

Fig. 7.16 Block diagram of earth station receiver (or downlink converter section)

(c) **Downlink Converter**: An earth station receiver has an input BPF, a LNA and a RF to IF down converter. Figure 7.16 exhibits the block diagram of a typical earth station receiver. The input BPF restricts the input noise power to the LNA, which normally uses a tunnel diode amplifier or a parametric amplifier. The RF to IF down converter consists of a mixer-BPF-combination, for converting the received RF signal to an IF frequency.

If three or more earth stations wish to communicate in a satellite communication system then any of the three methods of multiple accessing called frequency division multiple access (FDMA), time division multiple access (TDMA) and code division multiple access (CDMA) are required.

Advantages: Satellite communication has a number of positive features not readily available with other means of communication: (1) Since a very large area of the earth's surface is visible from a satellite, the satellite can provide line-of-sight coverage for a large number of users. (2) Thus, the satellite system provides numerous services, e.g. point-to-multipoint telecommunication links, which include telephone, TV, telegraphy, telex, FAX, videoconferencing, video text, digital transmission services, etc., to remote communities in sparsely populated areas like north-eastern states and hilly terrain like Ladakh, Himachal Pradesh, etc. (3) Finally, the satellite system is a better option for mass applications like search, rescue, mobile, meteorological and navigational purposes.

Thus, the world has been reduced to a *global village*, with the advent of the geostationary satellites and low altitude satellites like Iridium and Global Positioning System (GPS).

Disadvantages: (1) When communication is done through geostationary satellites, due to the large to and fro distance involved ($35,786 \times 2 = 71,572$ km), there is a large time delay of 250 ms between the transmission and reception of a signal.

(2) A satellite once launched and placed in its orbit, the malfunctions in the satellite is impossible to correct. (3) Therefore, initial cost involved is quite large due to the choice of highly reliable components. (4) It increases the amount of EM waves in the environment as pollution.

Communication Satellites: USA-based International Telecommunication Satellites Organization has launched and has been launching many INTELSATs (International Satellite) that provides telecommunication services to almost all countries throughout the world. Similarly, Indian Space Research Organization (ISRO) of India has been continuously launching INSATs (Indian Satellites) in a cheaper way which provide communication services to the Indian as well as other countries also. The telecommunication includes not only telephony but also telegraphy, TV, digital transmission services, telex, videoconferencing, video text, short message services (sms), etc.

7.8 Radar System

Radar is an acronym for radio detection and ranging. It is an electromagnetic system for the detection and location of objects, as it helps in extending human sensors. It can perform much better than human eyes, as it can see through fog, mist, darkness and is capable of measuring the distance of the object from the point of observation. The developments in radar have been tremendous and its applications are very wide. However, the principle with which the radar operates is much simpler than how human eyes perceive vision.

A simple radar system consists of an antenna, for transmitting electromagnetic signal and receiving the echo signal from the objects which intercepts the transmitted signal. Besides antenna, it also consists of a transmitter and a receiver to process the echo signal. Assuming that a pulse of electromagnetic signal is transmitted at time t_o, and that the echo signal is received at time t_1, the distance of the intercepting object (target) from the point of transmission is given by,

$$R = \frac{c(t_1 - t_0)}{2}$$

where $(t_1 - t_0)$ is the time taken by the transmitted signal to travel to the target and return, c is the velocity of the electromagnetic pulse, which is same as the velocity of light and R is the distance or range of the target. The above equation can be used to determine the range of the target.

Range Equation
The radar range equation relates the range, *i.e.* the distance between the radar and the target, to the characteristics of the transmitter, receiver, antenna, target and environment. A simple form of range equation is derived here (*i.e.* isotropically) assuming that the radar transmits signals in all the directions and so is the case of echo signal. Let the power of the signal transmitted from the radar be P_t watts. If the

gain of the omnidirectional transmitting antenna is G, then the power density of the signal intercepted by the target at a distance R is given by

$$\text{Power density from the antenna} = \frac{P_t G}{4\pi R^2} \tag{7.11}$$

The transmitted signal is intercepted by the target; however, the signal is not intercepted by the entire surface area of the target. Let us call the surface of the target that intercepts the transmitted signal as 'radar cross section of target σ'. The signal after being intercepted by the target is scattered in all directions and a portion of this signal which reaches the radar back is called the 'echo'. The power density of the echo signal P_{echo} which also spreads spherically is

$$\text{Power density of echo } P_{echo} = \frac{P_t G}{4\pi R^2} \cdot \frac{\sigma}{4\pi R^2} \tag{7.12}$$

The amount of power received by the radar from this echo signal is directly proportional to the effective aperture area A_e of the radar. Therefore, the further reduced power received (P_r) by the radar is

$$P_r = P_{echo} \cdot A_e = \frac{P_t G}{4\pi R^2} \cdot \frac{\sigma A_e}{4\pi R^2} = \frac{P_t G \sigma A_e}{(4\pi)^2 R^4} \tag{7.13}$$

From the Eq. (7.13) it is clear that the received power is reduced as the range of the target increases. If we define the maximum range R_{max} of a target as that range, beyond which the power of the received echo signal is negligibly small for detection, then

$$P_{r(min)} = \frac{P_t G \sigma A_e}{(4\pi)^2 R_{max}^4} \tag{7.14}$$

From the above equation, the maximum range of a target will be:

$$R_{max} = \left[\frac{P_t G \sigma A_e}{(4\pi)^2 P_{r(min)}} \right]^{1/4} \tag{7.15}$$

A Simple Radar System: The block diagram of a simple pulse radar is shown in Fig. 7.17 and it consists of a large number of blocks of circuits explained below:

(1) and (2) **Transmitter**: It consists of a high power oscillator and a modulator to generate a repetitive train of pulses.

      (3) **Duplexer**: A duplexer is used as a single antenna used for transmitting and receiving. The function or a duplexer is also to protect the receiver from damage caused by the high power of the transmitter. It main function is to channel the returned echo signals to the receiver and not to the transmitter.

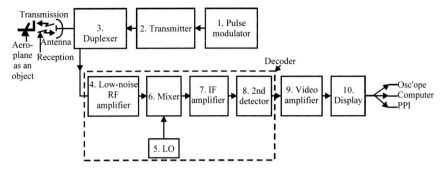

Fig. 7.17 Block diagram of simple pulse radar

(4) ***Low Noise RF Amplifier***: This forms the first stage of the super-heterodyne receiver, which amplify the received echo signals to an acceptable level before processing it.

(5) & (6) ***Local Oscillator (LO) and Mixer***: The mixer and local oscillator convert the RF signal to an intermediate frequency (IF of around 60 MHz, while in radio it was 455 kHz) (as done in a superheterodyne receivers in Fig. 7.12b) that is suitable for the receiver.

(7) & (8) ***IF Amplifier and Second Detector***: The output from the mixer is an intermediate frequency of around 60 MHz for a typical air-surveillance radar. As usual, IF amplifier increases the signal-to-noise ratio, *i.e.* it increases the signal power much above the noise power level.

In the transmitter section, the continuous signal from the oscillator is modulated to form a train of pulses. The second detector is used to extract this pulse modulation information.

(9) ***Video Amplifier***: It amplifies the signal before going to the display.

(10) ***Display***: The output from the second detector is amplified and is referred to as 'raw data' which is properly processed and applied to a display unit. The display unit can be anyone or more of the following:

(a) ***A-Scope***: A deflection-modulator CRT in which the vertical deflection is proportional to target echo strength and the horizontal coordinate is proportional to range.

(b) ***B-Scope***: An intensity-modulator display with azimuth angle indicated by the horizontal coordinate and range by the vertical coordinate;

(c) ***C-Scope***: An intensity-modulated display with azimuth angle indicated by the horizontal coordinate and elevation angle indicated by the vertical axis.

(d) ***PPI or Plan Position Indicator***: An intensity-modulated circular display sometimes called P-scope on which echo signals produced from reflecting objects are shown in plan position with range and azimuth angle displayed in polar coordinates, forming a map-like display.

(e) *RHI or Range-Height Indicator*: An intensity-modulated display with altitude as the vertical axis and range as the horizontal axis.

There are many other display devices with additional facilities like the display of alphanumeric characters, symbols, etc., which also may be chosen by a Radar station.

7.8.1 Types of Radars

A variety of radars have been developed and they are classified based on (*i*) the signal transmitted (*ii*) the target of radars and are discussed as follows.

1. *Continuous Wave (CW) Radar*: The radar transmitter transmits a continuous signal rather than a pulse one. The echo signal can be easily distinguished from the transmitted signal as there will be a shift in the frequency of the echo signal due to the Doppler effect. Also, the echo signal's strength is many times lower than the transmitted signal strength. The Doppler frequency shift is given by

$$\Delta f_d = \frac{2V_r f_o}{c} \qquad (7.16)$$

where

Δf_d Doppler frequency shift,
V_r relative velocity of the target with respect to radar,
f_o transmitted frequency, and
c velocity of propagation of electromagnetic signal, same as velocity of light = 3×10^8 m/s.

By using a CW radar, the relative velocity of a target can be determined, but the drawback is that the range of the target cannot be determined.

2. *Moving Target Indication (MTI) Radar*: The MTI radar transmits a train of electromagnetic pulses. This radar system is capable of distinguishing a moving target from the non-moving objects. Since there will not be any Doppler frequency shift from fixed targets, as there is no relative motion, they are easily identified from the moving target. The echo from the stationary objects, e.g. hills tower, etc., are called clutter and are removed. The echo from the moving target has varying frequencies due to Doppler effect.

3. ***Tracking Radar***: A tracking radar system works for a particular target and keeps
   facing it while it moves. It measures its coordinates and provides data from
   which the future course of the target can be predicted.

The tracking radar is similar to other radars except that the antenna beam of a
tracking radar is made to look at the target always. This is achieved by a ser-
vomechanism activated by error signals. The error signal is obtained using several
techniques, viz. sequential lobbing, conical scan and simultaneous lobbing.

Applications of Radar: Radar systems find their applications on the ground, in
the air, on the sea and in space. The ground-based radars have been used in the
detection, location and tracking of aircraft or space targets. The shipboard radars are
used to navigate the ships and to locate buoys, shorelines, other ships and observe
aircraft. The airborne radars on aircraft are used to detect land vehicles, ships and
other aircraft. However, the principal application of airborne radar is for mapping of
land storm avoidance and navigation. In space, e.g. in satellites, radar is used for
remote sensing purposes.

7.9   Data Transmission

7.9.1   Modem

The term modem is a acronym of the words **MOD**ulation and **DEM**odulation and
has these two functions. A telephone system, in which the signals are analog
carriers and the computer uses digital signals; therefore, it is necessary to have a
device to interface these two systems. This device is called **modem**. There are two
main functions that the modem performs. The computer's digital serial bit stream is
used by the modem to **modulate** a carrier signal suitable for transmission over the
phone lines. The modem at the receiving end **demodulates** the signals to recover
the signal to its original digital form.

There are two types of modem available: **asynchronous** and **synchronous**.
Asynchronous modem is generally used in telephone lines, wherein the data to be
transmitted is first divided into bytes. The bytes are sent bit by bit serially along
with an identification bit for '**start**' and '**stop**' purpose as shown in Fig. 7.18a. At

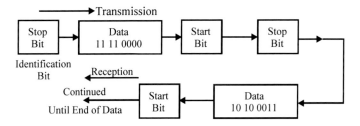

Fig. 7.18(a) Asynchronous modem

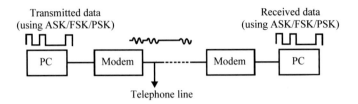

Fig. 7.18(b) Modem interfaced with telephone line and computer

the receiving end, the start and stop bit is identified and thereafter the data is recovered. The method of identification of start and stop bit is referred as hand-shaking process.

In synchronous modem, identification of start and stop bit is avoided. Data is directly sent and received by the use of careful timing-special type of clocking. This mechanism is used at the sending end modem and the receiving end modem also, and both the clocks are synchronized. The clock synchronization will be in such a way that the receiver recovers the correct data at the predetermined and pro-grammed interval.

Frequency-shift keying (FSK) method of modulation is used in low-speed asynchronous transmission for 300–1800 bits per second. The carrier frequency of the modem is shifted between two discrete frequencies in accordance with the logic levels of the digital signal. The higher frequency is referred as Logic 1, and the lower frequency is referred as Logic 0.

According to the requirement, other modulation like amplitude-shift keying (ASK) and phase-shift keying (PSK) may also be used. Quadrature amplitude modulation (QAM), which is a combination of ASK and PSK, is used to achieve for higher transmission rate.

If the modem is used to send and receive the signals in one direction only, then it is referred as 'simplex' mode operation. In 'half duplex' mode of operation, the communication is shared between transmission and reception. While transmitting, the receiver is switched off and while receiving the transmitter is switched off. In 'full duplex' operation, the modem can transmit and receive data simultaneously. Figure 7.18b is an example of using modem for transmitting the data signals interfaced with telephone lines and computer.

Functions of Modem (Transmitted end): Figure 7.19 shows a simplified block diagram of a modem used with RS-232 interconnection.

(i)   Take the data from RS-232 interface.
(ii) Convert the data (0's and 1's) into appropriate tones (modulation process).
(iii) Perform line control and signalling to the other end of the phone line.
(iv) Send dialling signals if this modem is designed to dial without the user present.
(v)   Have protection against line over voltage conditions and problems.

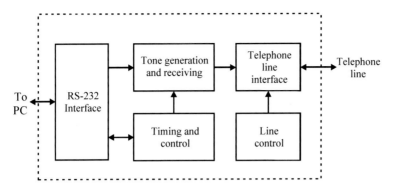

Fig. 7.19 Design of modem with RS-232 interface

Functions of Modem (Receiving end)

   (i)   Receive tone from the phone line.
  (ii)   Demodulate these tones into 1's and 0's.
 (iii)   Put demodulated signal into RS-232 format and connect to RS-232 interface.
 (iv)   Perform line control and signalling.
  (v)   Have protection against line over voltage problems.
 (vi)   Adapt its receiving system to variations in received noise, distortion, signal levels and other imperfections, so that the data can be recovered from the received signals.

7.9.2   Radio Transmission

The radio transmission system transmits radio waves, for which the signal is modulated using amplitude or frequency or phase modulation methods and transmitted to the communication medium. The received data is recovered after demodulating the signal at the receiver end. The general block diagram of radio transmitter and the receiver is shown in Fig. 7.20. The basic data called modulating signal is amplified. The carrier frequency wave is generated using high-frequency oscillator. Both the modulating and carrier waves are fed into the modulator. The modulated signal is further amplified using power amplifier and transmitted to the communication medium. The radio receiver is a device that picks up the signals from the receiver antenna. The tuning circuit selects the desired frequency range. The signal from the tuner is amplified and fed into the detector (demodulator) where the data is recovered and fed into the power amplifier. The amplified audio signal is then fed into the speaker.

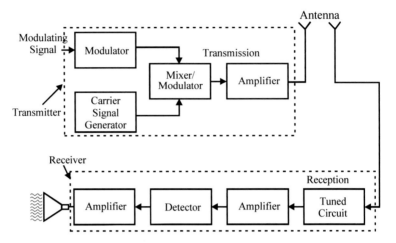

Fig. 7.20   The radio transmitter and receiver

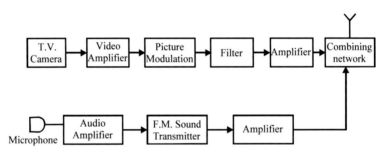

Fig. 7.21   T.V. transmitter

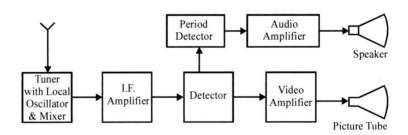

Fig. 7.22   T.V. receiver

7.9.3 Television Transmission

A television system is used for transmission and reception of visual and audio information by electronics medium. The general block diagram of T.V. transmitter and receiver is shown in Figs. 7.21 and 7.22. At the transmitter end, the camera converts the picture signal into electrical signals. The conversion is carried out by image 'orthicon' or 'vidicon' or 'plumbicon' methods operated using the principle of photoconduction. Video amplifier amplifies the electrical signal from the camera. The amplified picture signal is amplitude modulated, filtered and further amplified. The audio signal is separated from video signals. It consists of microphone, audio amplifier, transmitter and power amplifier. The amplified signal from microphone is applied to the modulator, which changes the radio frequency in accordance to the audio signal and produces the frequency-modulated wave. This signal is further amplified. Both the video and audio signals mixed in a combining network and transmitted through the same antenna.

In the tuner, the section function is accomplished, which selects one particular channel rejects other channels. The tuner section consists of local oscillator and mixer to convert the RF signal to an intermediate frequency signal (IF). This signal is amplified using IF amplifier. The signal is demodulated in detector section. The detected video and audio signals are amplified using separate video and audio amplifiers. Finally, the video signal is fed into the picture tube and the audio signal is fed into the speaker. The received signals are synchronized with the transmitted signals using synchronization pulses.

7.9.4 Microwave Transmission (Communication)

Electromagnetic waves in the frequency range of 1 GHz to 300 GHz are referred to as microwaves. As microwaves travel only on the line-of-sight (LOS) paths, the transmitter and receiver should be visible to each other. Hence, it is necessary to provide repeater stations in between the terminal stations at about 50 km intervals. As microwave communication offers a large transmission bandwidth, many thousands of telephone channels along with a few TV channels can be transmitted over

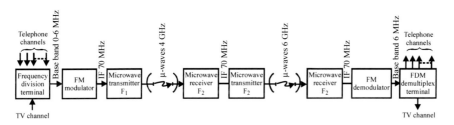

Fig. 7.23 Microwave transmission system

the same route using the same facilities. Normally, carrier frequencies in the 3–12 GHz range are used for microwave communication. The transmitter output powers can be low by using highly directional high gain antennas. Figure 7.23 shows the equipment needed to provide one channel of a ground-based microwave system.

It consists of two terminal stations with or more repeater stations. At the sending terminal, several thousand telephone channels and one or two television channel(s) are frequency multiplexed to form the baseband signal. The baseband signal is made to go for frequency modulation of an intermediate frequency (IF) carrier in the lower frequency range, which is then up converted to the microwave frequency of 4 GHz. This signal is amplified and fed through a directional transmitting antenna towards a repeater station at a distance of about 50 km.

At the repeater station, the signal is received on one antenna directed towards the originating station. The received signal is down converted to IF, amplified and up converted to a new frequency of 6 GHz. The frequency conversion is done so that the outgoing and incoming signals do not interfere with each other in the repeater stations. This signal is retransmitted towards the receiving terminal stations where it is down converted to the IF and demodulated to recover the baseband signal. This baseband signal is then demultiplexed to recover the individual telephone or television channel signals.

Presently, microwave communications are widely used for telephone networks, in broadcast and television systems and in several other communication application by services, railways, etc.

7.9.5 Optical Transmission

The principal motivations behind new transmission (*i.e.* a new communication) system are (*i*) to improve transmission fidelity, (*ii*) to increase the data rate (more information transmitted) and (*iii*) to increase the transmission distance between relay stations. Optical frequencies lie in the range 10^{14}–10^{15} Hz and the channel is fibre optics only. The laser information carrying capacity is greater than the microwave system by a factor of 10^5. An optical fibre can carry approximately 10 million TV channels. Optical fibre is used as a channel, which transmits, light signals. Optical fibre consists of a core material whose refractive index is greater than that of material around it known as cladding.

In optical communication, the transmitter is a light source which acts as a carrier wave through the fibre. This carrier wave modulated with digital signal by using ON and OFF technique called as 'on–off keying' or 'amplitude-shift keying'. In optical system, this is achieved by varying the source drive current directly causing a proportional change in optical power. The most common light source used are 'semiconductor laser diodes (SLD)' and 'light-emitting diodes (LED)'.

At the receiving end, a photodiode detector is used to convert the modulated light back to electrical signal. The photodiode current is directly proportional to the

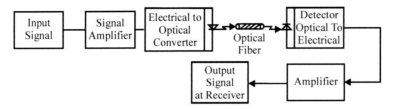

Fig. 7.24 Optical transmission system

Fig. 7.25 Refractive index profile and ray transmission in step index fibre

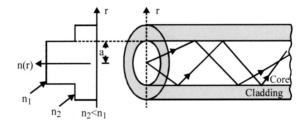

incident optical power. Normally, the output of the photodiode will be of small amplitude and the same is amplified according to the requirement of the user. The block diagram is shown in Fig. 7.24.

Optical Fibre: An optical fibre is a piece of very thin (hair-thin), highly pure glass, with an outside cladding of glass that is similar, but because of a slightly different chemical composition, has a lower refractive index. As shown in Fig. 7.25, the simplest optical fibre consists of a central cylindrical core of constant refractive index n_1 and a concentric cladding surrounding the core of slightly lower refractive index n_2. An optic fibre cable is quite similar in appearance to the coaxial cable system. This type of fibre is called *step index fibre*, whose core diameter is in the range of 2–200 μm, as the refractive index makes a step change at the core–cladding interface. The refractive index profile which gives the variation of refractive index with distance along the cross section of the fibre may be defined as

$$n(r) = n_1 \qquad \text{for } r < a \, (core)$$
$$= n_2 \qquad \text{for } r \geq a \, (cladding)$$

The refractive indices of core and cladding are related by the relative refractive index difference (Δ) between the core and the cladding by the relation

$$\Delta = \frac{n_1^2 - n_2^2}{2n_1^2}$$
$$= \frac{n_1 - n_2}{n_1}, \text{ where } n_1 \approx n_2$$

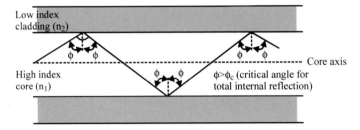

Fig. 7.26 Transmission of light ray in a perfect optical fibre

As the core and the cladding are normally made of glass or plastic, the refractive indices n_1 and n_2 lie around 1.5. Step index fibre may be used for multimode or single-mode propagation.

If a light ray travelling in the core of higher refractive index is incident at the core-cladding interface with an angle of incidence, with respect to the normal, greater than the critical angle, it will be reflected back into the originating dielectric medium, *i.e.* core, with high efficiency (around 99%). This phenomenon is known as ***total internal reflection***. In an optical fibre, transmission of light ray takes place by a series of total internal reflections at the core-cladding interface as shown in Fig. 7.26.

In the ***graded-index*** fibre, the refractive index gradually reduces from the centre to the outside of the fibre cross section. The graded-index fibres were initially easier to manufacture but lower attenuations are possible with step index fibres only, therefore the later are used nowadays.

Advantages of Optical Fibre Communication

(i) As the optical carrier in the range 10^{14}–10^{15} Hz is used, the system has enormous potential bandwidth.

(ii) Optical fibres have very small diameters, and hence, they are of small size and weight.

(iii) Optical fibres are fabricated from glass or a plastic polymer which are electrical insulators so that good electrical isolation in a hazardous environment is possible.

(iv) As optical fibre is a dielectric waveguide, it is free from electromagnetic interference (EMI), radio frequency interference (RFI), or switching transients giving electromagnetic pulses.

(v) As optical fibres do not radiate light, they provide a high degree of signal security and crosstalk between parallel fibres is avoided.

(vi) Optical fibre cables exhibit very low attenuation as compared with copper cables, making them suitable for long haul at least 1 km before a repeater is required in telecommunication applications.

(vii) Optical fibres are manufactured with very high tensile strengths and they are flexible, compact and extremely rugged.

(viii) They are highly reliable and easy to maintain.

(ix) The glass from which the optical fibres are made is derived from sand which is a natural resource. So, in comparison with copper conductors, optical fibres offer low-cost line communication.

(x) Optical fibres offer high tolerance to temperature extremes, corrosive liquids and corrosive gases and therefore have longer life span.

Applications

(i) In computers for linking them to peripheral devices.

(ii) Optical fibres are used in local network (LAN) and wide area networks (WAN) to link the computers in ring, star or bus topology.

(iii) Used is industrial electronics.

(iv) Used in telecommunication.

7.9.6   Integrated Services Digital Network (ISDN)

In general, the data from the computer is transferred to analog signal using modem and then transmitted to the communication system. The modem at the receiving end reconverts the analog data to digital form to be used by the other end computer. As an improvement to this system, a set of standard have been developed to provide a data transfer over a digital telephone network called **ISDN network**. In ISDN system, the signal originating from the digital source remains digital throughout the network. The ISDN is connected through communication channels. There are several communication channels available.

Channel A: 4 kHz analog channel.
Channel B: 64 Kbps digital channel.
Channel C: 8 kHz analog channel.
Channel D: 16 or 64 Kbps digital channel.
Channel E: 64 Kbps digital channel.
Channel H: 134 or 1536 or 1920 Kbps digital channel.

The standard channel combinations are: basic rate, primary rate and hybrid. Using any one of these standards, the ISDN is made to interface with the users.

The basic rate interface (BRI) uses two channels: (1) B-channels (Bearer) 64 K bits per second (kbps) for carrying the users voice, audio, video and data; and (2) D-channel (Delta) of 16 kbps capacities to exchange the signalling messages

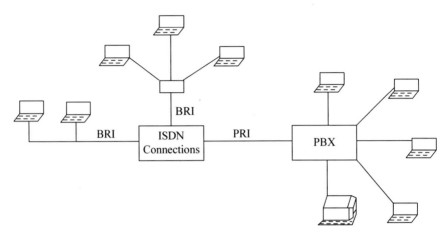

Fig. 7.27 ISDN interface

necessary to request services on the B channel. The basic rate interface is mostly used for residential applications with limited bandwidth.

The **primary rate interface (PRI)** delivers ISDN services used for digital PBX (private branch exchange), host computers and local area networks. It uses 23 or 30 B-channels at 64 Kbps for signalling and one D-channel for messaging at 64 kbps. The primary rate interface is normally used for large business applications with a private branch exchange (PBX). The general diagram is shown in Fig. 7.27.

The **hybrid rate interface (HRI)** uses one A-channel with 4 kHz analog and one *D*-channel with 16 Kbps digital.

Local area (LAN) networks and wide area networks (WAN) must fulfil the demand for the very high-speed data delivery. A new bandwidth-intensive services such as high definition television (HDTV) and ultra-high-speed computers are developed. In order to meet this need, a BROAD BAND ISDN (BISDN) has been developed and mostly used with fibre optic communication links, which is capable of transmitting at the speed ranging from 150 Mbps (megabits per second) to 2.5 Gbps (gigabits per second).

Questions

Fill in the blanks:

1. Communication is the process of _____ signals from one place to another.
2. The range of medium frequencies is _____.
3. Wavelength of ultra-high frequencies is _____.
4. In communication, the antenna dimensions should be of the order of _____ of the signal that is transmitted.

5. The total power in the modulated (AM) wave is expressed as _____.
6. The expression for current in amplitude modulation is _____.
7. Modulation index of FM is _____.
8. The master oscillator generates _____ carrier frequency.
9. The performance of AM receiver is improved by a technique known as _____.
10. The function of quantizer in receiver has to take a decision about _____ of a pulse.

Short questions

1. What is the difference between analog and digital signals?
2. What is the difference between ASK, FSK and PSK.
3. What is modulation?
4. What are the types of analog modulation?
5. Write the expression for modulation index in AM.
6. Write down the mathematical expression for a FM wave.
7. What are the advantages of FM over AM?
8. Compare FM and PM.
9. State sampling theorem.
10. Define the following terms:
    (a) Pulse amplitude modulation (PAM).
    (b) Pulse width modulation (PWM).
    (c) Pulse time modulation (PTM).
    (d) Pulse code modulation (PCM).

11. Describe briefly ASK, FSK and PSK.

Long questions

1. Explain briefly the need for modulation.
2. Draw the block diagram of communication system and explain its operation.
3. Explain amplitude and frequency modulation.
4. Describe briefly PAM, PCM, PPM and PDM.
5. Describe briefly about the data transmission.
6. Draw the block diagram arrangement of an AM transmitter and explain its operation.
7. Explain about the superheterodyne receiver with a neat block diagram.
8. Draw the block diagram of FM transmitter and explain its operation.
9. Draw the block diagram of FM receiver and explain its operation.
10. Explain about the radar system with a neat block diagram.
11. Draw the block diagram of modem and explain its operation.
12. Explain the satellite communication system with a neat block diagram.
13. Explain about the low power and high power AM transmitter with a neat block diagram.

Chapter 8
Basic Electronics Experiments and Lab Manual

Contents

8.1 Experiment 1: Breadboard and Component Mounting

Aim

To study about the breadboard and its connections.

Apparatus Requirement

Breadboard and electric/electronic components.

Theory

A circuit board that is used to make temporary circuits for experiments is called breadboard used, for test circuit designs. The electronic elements inside these electronic testing circuits can be interchanged by inserting terminals/leads into conducting (plated) holes and connecting it with the help of wires. The device has stripes of metal below the board that connects the conducting (plated) holes placed on the top of the board. The connections we make by wires on the breadboard are temporary, and the elements connected can be removed from the board and can be reused without any damage. Breadboards are generally used in electronics

© Springer Nature Singapore Pte Ltd. 2020
S. S. Srikant and P. K. Chaturvedi, *Basic Electronics Engineering*,
https://doi.org/10.1007/978-981-13-7414-2_8

engineering labs. Starting from tiny analog, digital circuits to big complicated CPUs, etc. can be tested with the help of this breadboard, at the initial stage, before making a full permanent PCBs.

The biggest advantage of using a breadboard is that these positions of the wires can be changed if they are placed in the wrong order. The diagram given below gives alphabets (A, B, C, D and a, b, c, to j) used in order to identify horizontal columns and numerals (1, 2, ...), in order to identify vertical columns. Every nth hole points of rows a, b, d and e are shorted by vertical copper strips conducting leads below the PCB. The same type of vertical shorting is done in another set of rows: e, f, g, h, i and j independently. For example, first holes of all the row points of rows 'a' to 'j' are shorted (i.e. a, b, c, ... j shorted). The same is true for second holes and so on. There are four more rows of holes, namely, A and B, above row 'a' and two rows C and D below row 'j'. In these four rows, each point is shorted horizontally by Cu strips below the PCB (Figs. 8.1 and 8.2).

As soon as the power is given to any point, e.g. 'a_2' of the line 'a' of the matrix, then all the points $a_2, b_2, c_2, d_2, \ldots j_2$ will be at some voltage as they are shorted.

As shown in Fig. 8.3, we can see how a resistor and LED connections are made on the breadboard. A 6 V battery is eventually attached to the LED light through a resistor because vertical line no. 7 in a, b, c, d, e is shorted and vertical line no. 3 in f, g, h, i, j shorted, which we are using here. If we replace the resistor with a new resistor having greater resistance, then we can see that the light intensity of LED becomes dimmer.

Mounting Components by Inserting Their Leads

The tiny sockets like holes are of 0.1 cm diameter arranged in a grid. The leads of the most passive or active elements can easily be pushed inside these holes. For inserting an IC, its dot identifier has to be kept at left side before placement (Fig. 8.4), then we use the isolated lines e and f across the gap for placing the IC. For connecting the holes, we require single-core plastic coated wires that have 0.6 mm diameter, because if we use non-standard wires it can damage the board.

For power supply, we normally use line 'A' and 'C' (see Fig. 8.4).

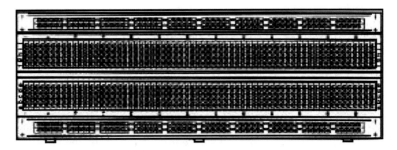

Fig. 8.1 Breadboard

Fig. 8.2 Breadboard in horizontal vertical position with pre-shorted lines by the printed Cu-strips below it is shown

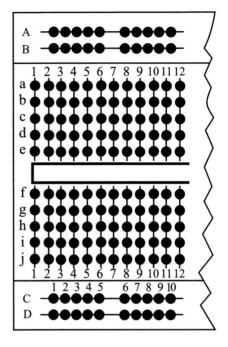

Fig. 8.3 LED with its power supply

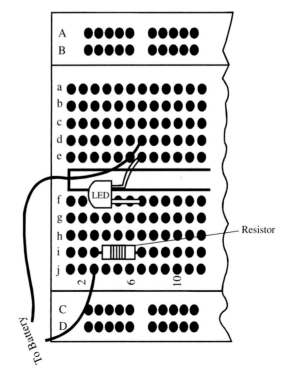

Fig. 8.4 Inserting an IC or
resistor

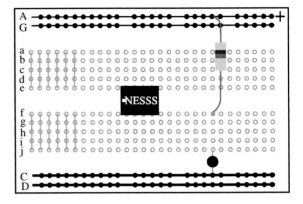

Inserting Components Parts into the Breadboard

In 3-D, the leads of the components into the breadboard are shown in Fig. 8.5, where we keep their pins straight and gently push into the holes. If the pins get bent, they can be straightened with a nose plier. The components should not touch each other. As shown in Fig. 8.5, the electrolytic capacitors have a positive and a negative electrode on which it indicates by a stripe with minus sign and sometimes arrowheads.

Thus, a student learns what a breadboard is and how to mount the components on it.

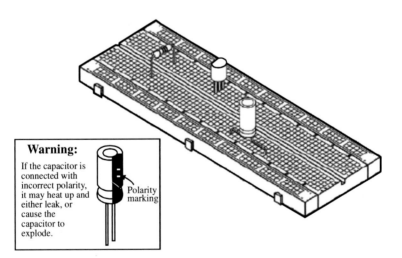

Warning:

If the capacitor is connected with incorrect polarity, it may heat up and either leak, or cause the capacitor to explode.

Polarity marking

Fig. 8.5 Inserting other components in breadboard

8.2   Experiment 2: Unknown Resistance Measurement

Aim

To study the resistor colour code system and to read the stated value of a resistor by interpreting the colour code indicated on the resistor.

Apparatus

Have some unknown resistances for study.

Theory

For finding the resistance value of a resistor, the colour bands on its body tell how much resistance it has. As shown in Fig. 8.6, there are four-band resistors and five-band resistors, the last band indicating the three types of tolerance as in Table 8.1. Table 8.2 gives the tolerance value, for all the colour codes of this last band, including those used in very precision circuits (Fig. 8.7).

Aid to Memory: The first letters of word to represent colour resistor code in Table 8.1:- **BBROYGBVGW** (**BBROY** of **G**reat **B**harat has **V**ery **G**ood **W**ife)

   Or (**B**etter **B**e **R**eady **O**r **Y**our **G**reat **B**ig **V**enture **G**oes **W**rong)

   Or (**BBROY** **G**ood **B**ye **V**ery **G**ood **W**ife).

   Figure 8.8 summarizes the four-band and five-band resistance reading with colour code along with Table 8.3.

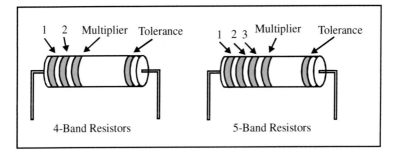

Fig. 8.6   Four-band and five-band resistors

Table 8.1 Four-band resistor code with tolerance

Colour	First Band	Second Band	3rd Band digit n as power of 10 as multiplier	Multiplier	4th Band (Tolerance colour code)
Black	///	0	0	$10^0 = 1$	///
Brown	1	1	1	$10^1 = 10$	///
Red	2	2	2	$10^2 = 100$	///
Orange	3	3	3	$10^3 = 1000$	///
Yellow	4	4	4	$10^4 = 10000$	///
Green	5	5	5	$10^5 = 100000$	///
Blue	6	6	6	$10^6 = 1000000$	///
Violet	7	7	7	$10^7 = 10000000$	///
Gray	8	8	8	$10^8 = 100000000$	///
White	9	9	9	$10^9 = 1000000000$	///
Gold	///	///	///	$10^{-1} = 0.1$	±5%
Silver	///	///	///	$10^{-2} = 0.01$	±10%
No Colour	///	///	///	///	±20%

Table 8.2 Resistor tolerance: last fourth band colour codes

Colour	Tolerance (%)
Gold	±5
Silver	±10
Gray	±0.05
Violet	±0.1
Blue	±0.25
Green	±0.5
Brown	±1
Red	±2

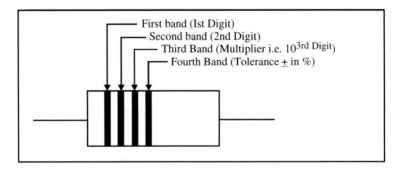

First band (Ist Digit)
Second band (2nd Digit)
Third Band (Multiplier i.e. $10^{3rd\ Digit}$)
Fourth Band (Tolerance ± in %)

Fig. 8.7 Method to read resistor values by colour codes in four-band resistor

Colour	Iˢᵗ Band	2ⁿᵈ Band	3ʳᵈ Band	Multiplier	Tolerance
Black	0	0	0	$10^0 = 1\Omega$	–
Brown	1	1	1	$10^1 = 10\Omega$	±1%
Red	2	2	2	$10^2 = 100\Omega$	±2%
Orange	3	3	3	$10^3 = 1K\Omega$	–
Yellow	4	4	4	$10^4 = 10K\Omega$	
Green	5	5	5	$10^5 = 100K\Omega$	±0.5%
Blue	6	6	6	$10^6 = 1M\Omega$	±0.25%
Violet	7	7	7	$10^7 = 10M\Omega$	±0.10%
Grey	8	8	8	–	±0.05%
White	9	9	9	–	–
Gold	–	–	–	0.1	±5%
Silver	–	–	–	0.01	±10%

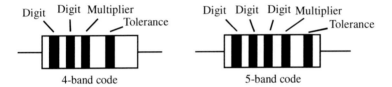

Skip this for 4-band resistors. Only for 5-band resistors!

Fig. 8.8 Four-band and five-band resistance measurement with colour code

Table 8.3 An example of calculating the given resistor value in Fig. 8.8

Band	Colour code	Numeric value	Noted value
(a) Top resistor in Fig. 8.8 (four-band system)			
1st band	Green	5	$56 \times 10^4 \pm 5\%$
2nd band	Blue	6	
3rd band	Yellow	10^4	
4th band	Gold	±5%	

The resistance value is 560 kΩ. The tolerance is ±5% = 560 kΩ ± 5%

In four-band system, the first band (green) shows the value 5, the second band (blue) shows 6, and the multiplier band or third band is yellow, means 4, which is multiplied by 10^4. Thus, the value is = $56 \times 10^4 \ \Omega \pm 5\% = 560$ kΩ ± 5%. The fourth band shows the gold, i.e. ±5% tolerance

(b) Bottom resistor in Fig. 8.8 (five-band system)			
First band	Red	2	$237 \times 10^0 \pm 1\%$
Second band	Orange	3	
Third band	Violet	7	
Fourth band	Black	0	
Fifth band	Brown	1	

The resistance value is 237 Ω. The tolerance is ±1% = 237 Ω ± 1%

In five-band system, the first band (red) shows the value 2, the second band (orange) shows 3, the third band (violet) shows 7, and the multiplier band or fourth band is black, means 0, which is multiplied by 10^0. The fifth band shows the brown, i.e. ±1% tolerance. Thus, the value is = $237 \times 10^0 \ \Omega \pm 1\% = 237 \ \Omega \pm 1\%$

Viva Questions

Q.1 Determine the value and tolerance of the two resistors (both four-band) as shown in the following tables.

Table A

Band	Colour code	Numeric value
1st band	Orange	
2nd band	Red	
3rd band	Orange	
4th band	Silver	
The resistance value is_____		The tolerance is _____

Table B

Band	Colour code	Numeric value
1st band	Gray	
2nd band	Violet	
3rd band	Orange	
4th band	Gold	
The resistance value is_____		The tolerance is _____

8.3 Experiment 3: *pn* Junction Diode Characteristics

Aim

To study and draw the forward and reverse bias I-V characteristics of a *pn* junction diode.

Apparatus Requirement

pn Diode,
Regulated Power supply (0–30 V),
Resistor 1 kΩ,
Ammeters (0–100 µA) and (0–100 mA),
Voltmeter (0–20 V),
Breadboard and connecting wires.

Theory

A *pn* junction diode conducts current only in one direction. The I-V characteristics of the diode are the curve between current as a function of variable voltage across the diode. Figure 8.9 shows the circuit connection for observing forward and reverse characteristics of a diode. When external voltage supply is zero, or the circuit is open, the circuit current is zero. In *pn* diode, when *p*-type end (Anode) is connected to positive (+ve) terminal and *n*-type end (cathode) connected to negative (−ve) terminal of the supply voltage, **it is known as forward bias**. The potential barrier formed by diffusion of opposite charges across the *pn* junction gets reduced when diode is in the forward-biased condition. At a forward voltage, greater than the potential barrier which is around 0.6 V or so in Si *pn* junction diode, the current starts flowing (in mA range) through the diode and also in the circuit. Then, the diode is called to be in **ON state**. The current starts at 0.6 V or so in Si diode and increases (which is always in mA range) with increase in forward voltage. When *n*-type (cathode end of *pn* junction) is connected to +ve terminal and *p*-type (Anode) is connected to the −ve terminal of the supply voltage, then it is **said to be in reverse bias** and the potential barrier formed by depletion of mobile charges, having to get attracted away from the junction on both sides by the +ve and −ve fields of voltage. The region across the junction gets depleted of charges (depletion region) and therefore, the junction resistance becomes very high and a very small current (minority current) flows in the circuit. The diode is then called to be in **OFF state**. This reverse bias current due to minority charge carriers is in µA range.

Circuit for Forward Bias and Reverse Bias Diode
See Fig. 8.9.

Expected I-V Characteristics of a Diode
See Fig. 8.10.

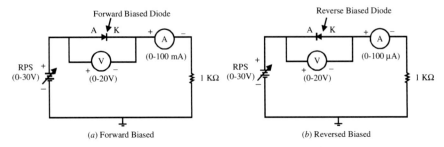

Fig. 8.9 Circuits for observing forward and reverse characteristics of a diode

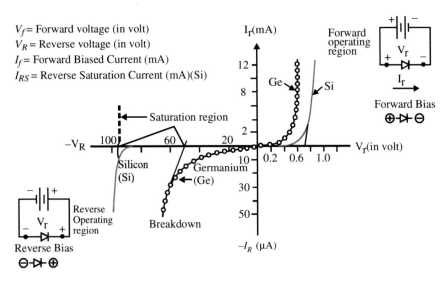

Fig. 8.10 Forward and reverse characteristics of a diode

Procedure
Case I: Forward-Biased Diode

1. Connections are as per the circuit diagram as in Fig. 8.9a.
2. The +ve terminal of RPS (Regulated Power Supply) is connected to the anode (*p* side) of the diode and its −ve terminal is connected to the (*n* side) cathode of the diode.
3. We switch on the power supply; keep increasing the input voltage from 0.2 V onwards in steps of 0.2 voltages steps.
4. The corresponding current flowing through the diode and voltage supplied across the diode for each step of the input voltage is noted and tabulated.
5. Therefore, we plot the I-V characteristic graph between forward voltage (in volt) and forward current (in mA).

Case II: Reverse-Biased Diode

1. Connections as per the circuit diagram as shown in Fig. 8.9b are made.
2. The +ve terminal of RPS is connected to the cathode (*n* side) of the diode and the −ve terminal of RPS is connected to the anode (*p* side), i.e. the diode is reverse biased now.
3. We switch on the power supply and keep increasing the input voltage (supply voltage) in steps of 0.5 voltage or so.
4. The corresponding current (in μA) flowing through the diode and the voltage across the diode is noted for each and every step of the input voltage and tabulated. As soon as the current goes to 100 μA, we stop increasing the voltage supply.

Finally, a graph is to be plotted between reverse voltage (in volt) and reverse current (in μA), showing a sharp increase of current beyond a certain voltage called breakdown voltage (see Fig. 8.10).

Observations (Sample)

The diode in forward bias

S. No.	Forward voltage (V)	Forward current (mA)
1	0.50	0.2
2	0.55	10
3	0.57	1.5
4	0.60	2.3
5	0.62	3.1
6	0.64	3.8
7	0.645	4.6
8	0.65	5.5
9	0.66	6.0
10	0.67	6.8

The Si diode in reverse bias

S. No.	Reverse voltage (V)	Reverse current (μA)
1	−0.76	−0.11
2	−1.50	−0.36
3	−2.15	−0.48
4	−2.86	−0.51
5	−3.56	−0.53
6	−4.28	−0.56
7	−5.00	−0.61
8	−5.8	−0.66
9	−6.50	−0.71
10	−7.16	−0.72
11	−65	−0.76
12	−80	−0.76

(continued)

(continued)

| The Si diode in reverse bias | | |
S. No.	Reverse voltage (V)	Reverse current (µA)
13	−85	−4.0
14	−90	−50.0

Precautions
1. All the connections to be as per Fig. 8.9.
2. Parallax error can be avoided by reading perpendicular to the metre while reading the analog metres.

Result
Thus, the forward and reverse bias characteristics for a *pn* diode was observed as per Fig. 8.10.

Viva Questions

Q.1 **Is there any relations between forward turn-on voltage of a diode with the energy band gap of material used?**

Ans. No, it is the built-in potential V_{BP}, which in reverse direction and formed by diffusion of carriers across the junction at zero bias (*n* from *n* region to *p* region and *p* from *p* region to *n* region). This amount of diffusion and the magnitude of V_{BP} depend on the doping densities of the *n* and *p* regions. Thus, a depletion region gets formed, associated with capacitance. When the diode is forward biased, then this bias first cancels the V_{BP} (being in reverse direction) and then only current starts flowing. Therefore, turn-on voltage is the measure of the built-in voltage (V_{BP}) formed at zero bias.

Q.2 **What is transition capacitance in a diode at reverse bias.**

Ans. Transition capacitance is formed in reverse-biased *pn* junction, where the carriers are taken/attracted away from the junction by the bias, thus making depletion layer as dielectric region with remaining part of *n*, *p* regions plates, as plate of a parallel plate capacitance, value of which will decrease with reverse bias voltage as depletion width (*d*) increases. Thus, $C_\tau \alpha 1/V_{Rev}$ and $C_\tau \alpha$ Doping, because

$$C_T = \varepsilon A/d$$

Q.3 **What is built-in potential or barrier potential at zero bias in a diode? Is there any capacitor formed here and what happens to it in forward bias.**

Ans. At zero bias, also the built-in potential is formed due to diffusion of carriers across the junction, thus forming depletion region here itself, associated with a capacitor. When we forward bias it, this capacitance increases as 'd' reduces, as the majority carriers of the two sides start crossing the junction. Thus, across the junction, charge is large (however, they keep recombining also) and distance between charges on the two sides 'd' is very very small. Therefore, capacitance is very very large and is called diffusion capacitance (C_D)

$$C_D = \tau I_D / \eta V_d$$

Q.4 **What is carrier lifetime in diodes?**

Ans. Here, τ is mean lifetime before recombination. I_D and V_d are diode current and diode voltage, while η is the generation–recombination factor. This factor depends upon semiconductor material and doping density.

Q.5 **Write something about the diffusion and transition capacitances.**

Ans. Thus, $C_D \gg C_T$ [Typically for a diode, $C_D = 20\,\mu\text{F}$ at 0.5 V and $C_T = 0.5\,\mu\text{F}$ at and −20 V]. In transistors at the *CB* junction, C_T is formed due to reverse bias and at the EB junction, C_D is formed due to forward bias. Using this concept of change of capacitance value with bias, varactor diode is made and is used very much in electronic circuits as tuning capacitor.

8.4   Experiment 4: Zener Diode Characteristics

Aim

To study and draw the static characteristics of a Zener diode and to find its voltage regulation.

Apparatus Requirement

Zener diode,
Regulated power supply (0–30 V),
Voltmeter (0–20 V),
Two Ammeter (0–50 μA),
Resistor (1 kΩ),
Breadboard and connecting wires.

Theory

A Zener diode is basically a heavily doped *pn* diode (in the range of doping of 10^{17} to 10^{18}/cc, below which it is avalanche diode, i.e. ordinary diode and above it a diode becomes degenerate, making it a tunnel diode). The Zener diode is specially made to operate in the breakdown region. A *pn* junction diode normally does not conduct when reverse biased, but if the reverse bias is increased, then beyond a particular voltage it starts conducting heavily. This voltage is called breakdown voltage and in case of Zener diode as Zener breakdown (V_z) which is very sharp in terms of increase of current even by increase of 0.1 V. High current through the diode can permanently damage the device to avoid high current, and we connect a resistor in series with the Zener diode. Once the diode starts conducting, it maintains almost constant voltage across the terminals, i.e. it has very low dynamic resistance (dV/dI) and because of its sharp breakdown, the Zener diode is used in voltageregulators (Refer Fig. 2.10 for comparison with Fig. 8.11b). The Zener diode regulates the voltage keeping across it fixed at V_z even when the supply voltage V_{AB}, crosses V_z. The extra voltage is absorbed by R (see Fig. 8.11b).

Circuit Diagram

See Fig. 8.11a.

I-V Characteristics of a Zener Diode

See Fig. 8.11b.

Procedure

Static Characteristics

1. Connections of the circuit are done as per diagram shown in Fig. 8.11a.
2. The regulated power supply voltage is increased in steps of 0.05 V.
3. The Zener current (I_Z) and the Zener voltage (V_Z) are observed and noted in tabular form, with forward and reverse biasing given to it.
4. Then, finally a graph is plotted between Zener current (I_Z) and Zener voltage (V_Z) (Fig. 8.11b).

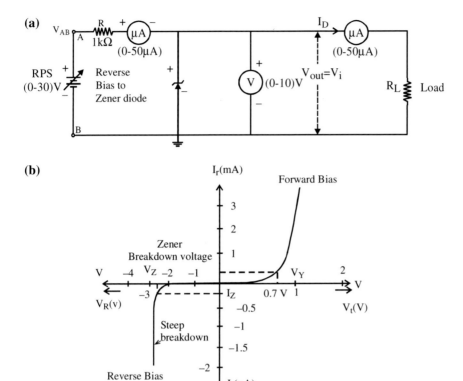

Fig. 8.11 **a** Circuit for Zener diode characteristics (Zener diode is used in reverse bias only). **b** I-V characteristics of a typical Zener diode—above circuit shows reverse bias, which can be reversed for forward bias

Regulation Characteristics

1. The voltage regulation of any device is normally expressed in percentage, given by the following formula:
2. $\text{Reg} = [(V_{NL} - V_{FL})/V_{FL}] \times 100$, where $V_{NL} =$ Voltage across the diode, without any load connected, i.e. $R_L = \infty$ and $V_{FL} =$ Voltage across the diode, when full load is connected (i.e. R_L) so that $I_0 R_L = V_z$.
3. The connection is made as per the circuit diagram shown in Fig. 8.11a.
4. The diode is placed in full-load condition with R_L connected and the Zener voltage (V_Z), Zener current (I_Z), load current (I_L) are measured by increasing supply voltage in steps of 0.05 V.
5. This above step is repeated by decreasing the value of the load, i.e. R_L in steps.
6. The readings are tabulated and plotted in graph.
7. Finally, the percentage regulation is calculated using the above formula.

Observations

Forward bias

S. No.	Forward diode voltage V_f (V)	Forward current I_f (mA)
1	0.50	0
2	0.55	0.2
3	0.60	0.7
4	0.65	1.2
5	0.68	1.8
6	0.70	2.3
7	0.73	2.8
8	0.75	3.2
9	0.76	3.8
10	0.77	4.8
11	0.80	5.0

Reverse bias

S. No.	Reverse diode voltage V_r (V)	Reverse current I_r (μA)
1	−0.5	0
2	−1.0	0
3	−1.60	0
4	−2	0
5	−2.6	0
6	−2.7	−0.15
7	−2.9	−0.22
8	−3.0	−0.55
9	−3.1	−1.05
10	−3.2	−1.55
11	−3.2	2.00

Load regulation at 5 V supply

Load (R_L)		Reg. (%)
Full	(1 kΩ)	0
¾ Load	(750 Ω)	0.1%
½, i.e. Half load	(500 Ω)	0.2%
¼, i.e. Quarter load	(250 Ω)	0.5%

Precautions

1. The terminals of the Zener diode should be properly connected, i.e. After identifying the p and n sides.
2. When we determine the load regulation, load should never be shorted, but with different R_L.
3. We have to be ensured that the applied voltages and especially currents do not exceed the ratings of the diode, given with that specific diode.

Result

1. Thus, the static characteristics of Zener diode were obtained and drawn as shown in Fig. 8.11b.
2. Percentage regulation of Zener diode was also calculated.

Viva Questions

Q.1   What are the differences between avalanche diode, Zener diode, and tunnel diode in terms of (a) I-V characteristics in forward and reverse bias, and (b) reverse bias breakdown voltage range and current range.

Q.2   Can we use avalanche diode as voltageregulator and why?

Q.3   What is the knee voltage in a diode?

8.5   Experiment 5: I-V Characteristics of LED

Aim

To study the I-V characteristics of LED.

Apparatus Requirement

Light-emitting diode (LED),
Voltmeter (0–20 V),
Ammeter (0–100 mA) and (0–100 μA),
Resistor (1 kΩ),
dc power supply (0–30 V) and
Breadboard and connecting wire.

Procedure

Connections are made as per circuit requirement shown in Fig. 8.12a. In forward bias, the bias voltage is varied in steps of 0.01 V and corresponding value is noted. In reverse bias, the bias voltage is varied in steps of 1 V and the corresponding ammeter reading is noted. The graph is plotted between forward voltage and forward current or reverse voltage and reverse current.

Theory

LED is a *pn* junction device made of GaP, GaN, SiC etc., which emits visible light when forward biased by a phenomenon called electroluminescence. In all semiconductor *pn* junctions, some of the energy will be radiated as heat and some in the form of photon ($\Delta E = h\nu$). In silicon and germanium, most of the energy is given out in the form of heat, i.e. infrared. (See Chap. 4 for detailed theory.)

When LED is forward biased, the electron and holes move towards the junction and recombination takes place. As a result of recombination, the electrons lying in the conduction bands of *n* region fall into the hole lying in the valance band of a *p* region. The difference, i.e. band gap $\Delta E = (E_C - E_V) = h\nu$, corresponds to optical light frequency in material like GaP, GaN, SiC, etc.

The brightness of the emitted light is directly proportional to forward bias current.

Circuit

See Fig. 8.12a.

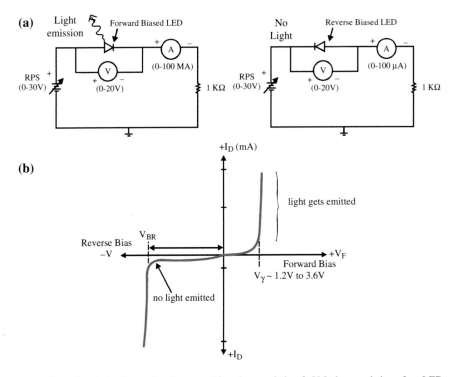

Fig. 8.12 a Circuit for forward and reverse bias characteristics. **b** V-I characteristics of an LED

Observations

Forward bias LED

S. No.	RPS (V) LED (V)	Voltage across diode (V)	Forward current I_D (mA)
1	0.5	0.5	0 No light
2	1	1	0 No light
3	1.5	1.5	0 No light
4	2	1.7	0.2 No light
5	2.5	1.7	0.7 Emits light
6	3	1.7	1.2 Emits light
7	3.5	1.8	1.7 Emits light
8	4	1.8	2.1 Emits light
9	4.5	1.8	2.6 Emits light
10	5	1.8	3.1 Emits light
11	5.5	1.8	3.6 Emits light
12	6	1.8	4.1 Emits light

Reverse bias LED

S. No.	RPS (V)	Voltage across LED (V)	Reverse current $-I_D$ (μA)
1	0.5	−0.5	0 No light emitted
2	1	−1	0 No light emitted
3	1.5	−1.5	−0.1 No light emitted
4	2	−2	−0.1 No light emitted
5	2.8	−2.8	−0.2 No light emitted
6	3	−3	−0.3 No light emitted
7	3.5	−3.5	−0.3 No light emitted
8	4	−4	−0.4 No light emitted
9	4.5	−4.5	−0.4 No light emitted
10	5	−5	−0.5 No light emitted
11	5.6	−5.6	−0.5 No light emitted
12	6	−6	−0.6 No light emitted

Expected I-V Characteristics of an LED
See Fig. 8.12b.

Results
Thus, the forward and reverse bias characteristics of LED have been studied.

Viva Questions
Q.1   The optical light falls between $\lambda = 0.73$ μm (IR) and $\lambda = 0.45$ μm (UV), does it mean that the band of LED materials should correspond to them with their band gaps? Is there a single material for making LED that could emit all the colours together?

Q.2   Why Si, Ge, InSb and GaAs cannot be used for making an LED.

Q.3   List the materials used to make LED.

Q.4   What is the working principle of LED?

Q.5   Prove that Si and Ge will emit IR but not in optical radiation range.

[Hint: Use the relation λ (μm) $= 1.24/E_g$ (eV)]

8.6   Experiment 6: Characteristics of Light-Dependent Resistor, Photodiode, Phototransistor

Aim
To determine the V-I characteristics of light-dependent resistor (LDR), photodiode and phototransistor.

Apparatus Requirement
Photodiode,
Phototransistor,
Light-dependent resistor (LDR),
Resistor 1 Ω,
Voltmeter (0–10 V),
dc power (0–30 V),
Ammeter (0–10 mA), (0–500 μA),
Lamp (100 W), and
Breadboard and connecting wires.

Introduction

Light-Dependent Resistor (LDR)
The photoresistor or light-dependent resistor (LDR) is made from high resistance semiconductor cadmium sulphide (CdS) materials. The resistance of LDR decreases on increasing the incident light intensity. It is also called as a photoconductor. If light is falling on the LDR, then the photons of light get absorbed by the semiconductor, which gives bound electrons of the CdS material enough energy to jump into the conduction band. The resulting free electron (and its hole partner) conducts electricity and hence lowers the resistance.

Photodiode Detector
A silicon photodiode is a semiconductor light intensity detector that consists of a shallow *p* impurity diffused *pn* junction. When the top surface of (*p*) is illuminated, photons of light having IR also penetrate into the very thin *p* layer (transparent) and then the photon energy reaches the *pn* junction generating electron hole pairs (EHP), as the bandgap of Si ($E_g = h\nu$) falls in IR frequency (ν) range. We know that in a *pn* junction, when reverse biased, only minority current, i.e. leakage current is there. With light falling on *pn* junction, the EHP generated increases the reverse bias leakage current. This increase is proportional to the light intensity, and therefore it becomes the measure of light intensity in photodiode (Fig. 4.10). In Solar cell diodes also light generate EHP, with 'e' getting attracted/accumulated on

'*p*' side, while '*h*' on '*n*' side. By shorting externally the current inside flows from '*n*' to '*p*', i.e. reverse of normal forward bias.

Phototransistor Detector

Phototransistor is like a photodiode for detecting light waves. The phototransistors, like transistor, are designed to be like a fast switch with very thin *p* region base and are also used for light-wave communications or infrared sensors. The most common form of phototransistor is the *npn* transistor, where we keep the base lead open, i.e. '*p*' is not connected. Base is connected to the emitter with a small resistor in some circuits. Light or photons enter the base from the collector–base junction diode side (For this, there is a transparent window in the packaging of the phototransistor). This creates *e-h* pair causing current flow, which replaces the base–emitter current of normal transistors, and hence gets amplified. Thus, a phototransistor is better light detector than photodiode, as the transistor acts as an amplifier of the current also, increasing collector–base current and thus increases the sensitivity of the detector.

Circuit Diagram

See Fig. 8.13.

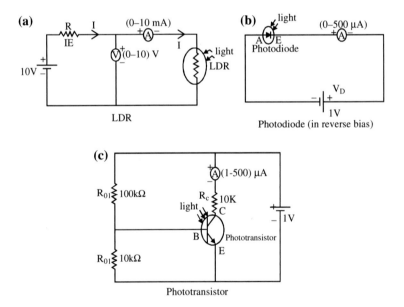

Fig. 8.13 Circuit for characteristics of **a** LDR, **b** photodiode and **c** phototransistor

Observations

LDR

Distance (cm)	Voltage across voltmeter (V)	Current across ammeter (mA)	Resistance (kΩ)
0	3	9	0.333
2	5.8	6	0.967
3	6.8	3.8	1.789
4	7.2	3.2	2.25
6	7.8	2.4	3.25

S. No.	Distance of light source from photodiode/phototransistor (in cm)	Photodiode Current (μA)	Phototransistor Current (μA)
1	0	180	450
2	2	80	330
3	4	50	200
4	6	20	100
5	8	10	50
6	10	2	5
7	200	1 (dark current)	2

Characteristics of LDR, Photodiode, Phototransistor

1. The dark resistance of an LDR (i.e. in dark) is very less as compared to the reverse bias resistance in photodiodes, because of which it may consume comparatively more power than its semiconductor counterparts, e.g. phototransistor, etc.
2. Phototransistors made of silicon are capable of handling voltages up to 1,000 V.
3. Normally, phototransistors are also more vulnerable to surges and spikes of electricity as well as electromagnetic energy, while the photodiodes are rugged with LDR most rugged one.
4. The silicon phototransistor has maximum sensitivity in the infrared (around a wavelength of 940 nm), which is typical for silicon photodiodes also.
5. If the lab has a lux metre, then that can be kept near the light detector (LDR photodiode/phototransistor) and check the sensitivity or calibrate it as well.
6. Photoresistors are much less light-sensitive devices than photodiodes or phototransistors, with phototransistor the most sensitive semiconductor device. The photoresistor is a passive component and does not have a *pn* junction. The photoresistivity of any photoresistor varied widely with ambient temperature, making them unsuitable for applications requiring precise measurement of sensitivity to light.
7. One more drawback of photoresistors is that it exhibits a certain degree of latency between exposure to light and the subsequent decrease in resistance, usually around 10 ms. This lag time when going from light to dark environments is much longer, as long as one second.

Precautions

1. In these experiments, we have to be cautious not to exceed the ratings of the diode, as this may damage the diode.
2. Voltmeter and ammeter should be connected as per polarities, shown in the circuit diagram.
3. Power supply to be switched **ON** only after we have checked the connections as per the circuit diagram.

Procedure

LDR

1. Circuit to be connected as shown in Fig. 8.13a.
2. Light source should be kept at a distance of 2 cm and switch it ON, with the light falling on the LDR.
3. Current and voltage in ammeter and voltmeter are to be noted in a table.
4. Now we increase the distance of the light source and again note the I and V.
5. Finally, we plot the graph between R as calculated from observed I-V and distance of light source.

Photodiode

1. Circuit to be connected is shown in Fig. 8.13b.
2. Distance between the bulb and photodiode needs to be kept 2 cm or so initially.
3. Keep the light bulb at fixed voltage. Only vary the voltage of the diode in steps of 1 V and note the diode current I_r.
4. Repeat the above procedure for $V_L = 2\,V, 4\,V$, etc.
5. Finally, plot the graph: V_d versus I_r for constant light source.

Phototransistor

1. Circuit to be connected as shown in Fig. 8.13c.
2. We repeat the procedure as that in the photodiode.

Note: If we have lux meter in the lab; then all these three devices can be calibrated. Then the light intensity (in lux) can be plotted as a function of resistance for LDR, current in case photodiode and phototransistor.

Expected Characteristics/Plot for LDR, Photodiode and Phototransistor
See Fig. 8.14.

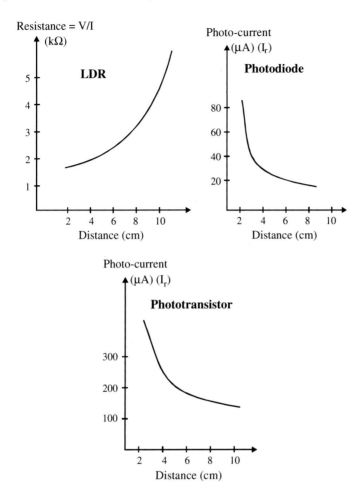

Fig. 8.14 A sample characteristic plot of LDR, photodiode and phototransistor for change of distance with light source, i.e. change of intensity

Result
Characteristics of LDR, photodiode and phototransistor were plotted and studied.

Viva Questions

Q.1   What are the principles of operation of LDR, photodiode and phototransistor?

Q.2   Give the applications of LDR, photodiode and phototransistor?

Q.3   Write the difference between photodiode and phototransistor.

Q.4   Photogeneration and recombination are opposite in action. Justify.

Q.5   What do we mean by dark current in photodiode?

Q.6   A photodiode can be operated in forward bias condition? Justify.

Q.7 Light has to fall on collector–base junction in case of phototransistor. Why?

Q.8 What are direct and indirect semiconductors and which is used as opto-electronic device? Why?

Q.9 If the distance of light source is increased in case of LDR, photodiode and phototransistor, what parameters change in each and how?

Q.10 What is Lux?

Q.11 What is optical range of wavelength?

Q.12 If we need to detect wavelengths from 400 to 1700 nm, which material must be used to manufacture phototransistor? Can a single material be used for the whole range?

8.7 Experiment 7: Half-Wave Rectifier

Objective

1. Plotting the output waveform of the half-wave rectifier.
2. Calculating the ripple factor for half-wave rectifier using the formulae.
3. Computing the efficiency, $V_r(pp)$, V_{dc} for half-wave rectifier.

Apparatus Requirement
Voltmeter (0–10 V),
Ammeter (0–50 mA) and (0–100 μA),
One transformer 6-0-6 V, 500 mA, 1 A rating,
One resistance 470 Ω, 10% tolerance, 1/2 W rating,
One capacitor 470 μF, and
One diode.

Introduction
Rectifier is a device for converting a sinusoidal input waveform (*ac*) into a unidi-rectional (*dc*) waveform with non-zero average component. A half-wave rectifier uses only one-half of the waveform (upper or lower part). A half-wave rectifier circuit with a resistive load is shown in the circuit Fig. 8.15. In positive half cycle, diode D is forward biased and conducts; therefore, the output voltage is the same as the input voltage. In the negative half cycle, diode D is reverse biased, and therefore output voltage becomes zero. A filter is needed between the rectifier and load for attenuating the voltage ripple component over the *dc* voltage output. A filter is simply a capacitor connected from the rectifier output to ground, which quickly charges at the beginning of a cycle and slowly discharges through RL after the positive peak of the input voltage, when no output is coming and thus it reduces the variation in output *dc* variation. The variation in the output voltage is due to charging amplitude of the *ac* input voltage with time, being sinusoidal and is called ripple voltage. Ripple is undesirable; therefore the smaller the ripple, the better is the filtering action (Fig. 8.15).

$$V_{rms} = \frac{V_{max}}{2}$$

Circuit Diagram of Half-Wave Rectifier
To measure of the effectiveness of a rectifier circuit, a formula called ripple factor is used to define a ratio of RMS value of *ac* component to the amplitude of the *dc* component of the rectifier output.

Theoretical Calculations for Ripple Factor
Case A: Without Filter

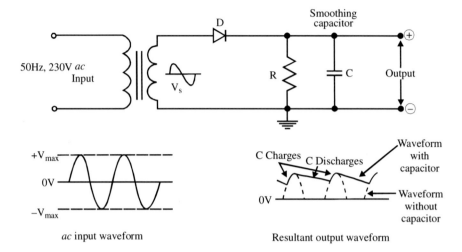

Fig. 8.15 Circuit for half-wave rectifier

$$V_{dc} = \frac{V_{max}}{\pi}$$

$$\text{Ripple factor (Theoretical)} = \sqrt{\left(\frac{V_{rms}}{V_{dc}}\right)^2 - 1} = 1.21$$

$$\text{Ripple Factor (practical)} = \frac{V_{ac}}{V_{dc}}$$

$$where \qquad V_{ac} = \sqrt{\left(V_{rms}^2 - V_{dc}^2\right)}$$

Case B: With RC Filter

$$\text{Ripple Factor (Theoretical)} = r = \frac{1}{2\sqrt{3}fCR}$$

$$V_{dc} = V_{max} - \frac{V_{r(p-p)}}{2}; V_{ac} = \frac{V_{r(p-p)}}{2\sqrt{3}}$$

where $V_r(p - p)$ is the ripple peak voltage after filter, at the output.

$$\text{Ripple Factor (practical)} = \gamma = \frac{V_{ac}}{V_{dc}}$$

$$\text{Percentage Regulation \%} = \frac{V_{NL} - V_{FL}}{V_{FL}} * 100$$

V_{NL} = No-load dc voltage at the output without connecting the load (when current is very small).

V_{FL} = Full-load dc voltage at the output with the load connected where current supplied increases.

$$\text{Efficiency} (\%) = \eta = \frac{P_{DC}}{P_{AC}} \times 100$$

$$\text{where } P_{AC} = V_{rms}^2 / R_L$$

$$P_{DC} = V_{dc}^2 / R_L$$

For reducing the ripple factor, we have to increase the load capacitance.

Characteristics of Half-Wave Rectifier

1. Power output as well as rectification efficiency in half-wave rectifier is quite low, due to the fact that power is delivered only during one-half cycle of the input ac voltage, the other half remains unused.
2. The output current in the load, in addition to dc component, contains ac components of basic frequency equal to that of the input voltage frequency. Ripple factor is high and an elaborate filtering is, therefore, required to give steady dc output.
3. The transformer utilization factor is low, as current is drawn in one-half cycle only.

Precautions

1. In the experiment, we have to be careful not to exceed the ratings of the diode as this may damage the diode.
2. The CRO is to be connected using probes properly as in the circuit diagram.
3. Switch **ON** the power supply only after having checked the circuit as per the circuit diagram.

Experiments

1. Connections are normally given as per the circuit diagram without capacitor and the student has to connect it.
2. The rectified output voltage is given to the CRO for measuring the time period and amplitude of the output waveforms (Fig. 8.16a, b).
3. Connect the capacitor in parallel with load resistor and then again note down the amplitude and time period of the output waveform, with this filter.
4. Also measure the amplitude and time period of the transformer primary winding (input waveform) by connecting CRO (Fig. 8.16) to the input side.
5. On the graph sheet, we now plot the input and output voltages without filter and with filter in y-axis and time in x-axis with appropriate scales.
6. Ripple factor can be calculated now.

Observations in CRO (Sample)

Parameters	(a) Input waveform	Output waveform (without filter)	Output waveform (with filter)
Peak amplitude (V)	3.30 × 100 V = 330 V	1.7 × 5 = 8.5 V	1.6 × 5 = 8.0 V
Time period (ms) [T]	4.9 × 10 ms = 49 ms (peak to peak)	4.9 × 10 ms = 49 ms (peak to peak)	Growing = 0.9 × 10 = 9 ms Drooping = 4.0 × 10 = 40 ms Total (T) = 49 ms
Frequency computed (Hz) [=2/T]	40.8 Hz	40.8 Hz	40.8 Hz
Ripple factor	=1.5/8 = 0.1875		
Regulation	{(8.5 − 8.0)/8.0} × 100 = 6.25%		

Plot of Half-Wave Rectifier
See Fig. 8.16.

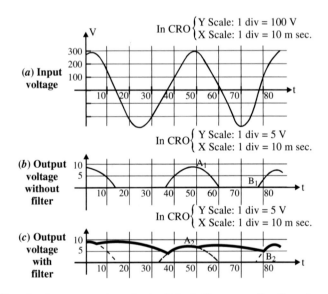

Fig. 8.16 Plot of voltage versus time circuit of the half-wave rectifier for **a** input waveform, **b** output waveform (without filter), **c** output waveform (with filter)

Result

The rectified output voltage of half-wave rectifier circuit as shown in Fig. 8.16 was observed, and the calculated value of ripple factor is 0.18, regulation = 6.25%.

Viva Questions

Q.1   **What is the purpose of a rectifier?**

Ans.   Rectifier is used to convert an *ac* signal to a *dc* signal.

Q.2   **Why filter is used in a rectifier?**

Ans.   They are used to remove or reduce the ripple in the voltage waveform.

Q.3   **What are the advantages of half-wave rectifier?**

Ans.   It is simple, cheap and easy to conduct.

Q.4   **Define ripple factor and efficiency. State the ideal value.**

Ans.   The ripple factor is a measure of fluctuating component in the output voltage. Ripple factor of half-wave rectifier with no filter is 1.21. Value of ideal condition efficiency is the ratio of *dc* power delivered to the load and to the *ac* power input given to the rectifier signal circuit. For half-wave, $\eta = 40.6\%$ in ideal condition.

8.8   Experiment 8: Full-Wave Centre-Tapped Rectifier

Aim

To study the working of a full-wave centre-tapped rectifier with and without filter and also measure its parameters.

Apparatus Requirement

Voltmeter (0–10 V),
Ammeter (0–50 mA) and (0–100 μA),
One transformer 6-0-6 V, 500 mA, 1 A rating,
One resistance 470 Ω, 10% tolerance, 1/2 W rating,
One capacitor 470 μF, and
Two diodes.

Introduction

In general, a rectifier converts sinusoidal input waveform (*ac*) into a unidirectional waveform (*dc*) with non-zero average component.

A full-wave centre-tapped rectifier with a resistive load is shown in Fig. 8.17, which consists of two half-wave rectifiers connected to a common load. The centre-tapped transformer supplies the two diodes (D1 and D2) with sinusoidal input voltages that are equal in magnitude but opposite in phase. One rectifies during positive half cycle of the input and the other rectifying the negative half cycle. During input positive half cycle, diode D1 is ON (passing on the +ve/upper half of the wave to the load) and diode D2 is OFF. During negative half cycle, diode D1 is OFF and diode D2 is ON passing on the lower half of the wave to the load in the same +ve voltage side (Fig. 8.17). Here, peak inverse voltage (PIV) is the maximum voltage that a diode can withstand when it is reverse biased. Peak inverse voltage for a full-wave rectifier is $2V_m$, because the entire secondary voltage appears across the diode, which is not conducting during the +ve and −ve cycles of the wave.

The full-wave rectifier's output also contains both *ac* and *dc* components. The applications normally cannot tolerate a high-value ripple, necessitating further processing of the rectified output. The undesirable *ac* components called ripple can be minimized using filters.

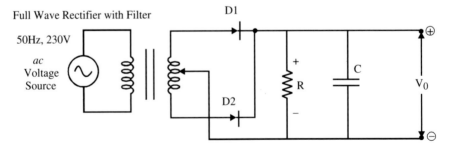

Fig. 8.17 Full-wave centre-tapped rectifier circuit with RC filter

Ripple Factor
Ripple factor of a rectifier is the ratio of the effective value of *ac* components to the average *dc* value, denoted by the symbol 'γ'.

$$\gamma = \frac{V_{ac}}{V_{dc}}, (\gamma = 0.48)$$

Efficiency
The ratio of square of the output *dc* voltage to the square of input *ac* voltage is defined as efficiency (η).

Circuit Diagram
See Fig. 8.17.

$$\eta = \frac{dc \text{ output power}}{ac \text{ input power}} = \frac{P_{dc}}{P_{ac}} = \frac{(V_{dc})^2}{(V_{ac})^2}$$

$\eta = 81.2\%$ (maximum efficiency of a Full Wave Rectifier)

Percentage Regulation
Percentage regulation is a measure of the ratio of the difference of *dc* output voltage without load and with load to the *dc* voltage at full load.

$$\text{Percentage regulation} = \left(\frac{V_{NL} - V_{FL}}{V_{FL}}\right) \times 100\%$$

Here V_{NL} = No-load voltage across load resistance in no-load condition, when minimum current flows through it.

V_{FL} = Full-load voltage across load resistance in full-load condition, when maximum current flows through it.

For a full-wave rectifier, the percentage regulation is close to 0% or very small.

Peak Inverse Voltage (PIV)
The maximum reverse voltage that the diode has to withstand (as explained earlier) is PIV.

Normally, **PIV $= 2V_m$**, where V_m is the peak value of *ac* voltage on the diode.

Transformer Utilization Factor
The transformer utilization factor (TUF) is defined as the ratio of *dc* power delivered to the load P_{DC} to the *ac* rating of the transformer secondary P_{AC}.

$$\text{TUF} = P_{DC}/P_{AC}$$

TUF can be used to determine the rating of a transformer secondary and measured by considering the primary and the secondary winding separately. Normally, it has a value of 0.693 or so.

Theoretical Calculations

$$V_{rms} = \frac{V_m}{\sqrt{2}} \text{ without filter}$$

$$V_{ac} = \sqrt{(V_{rms}^2) - V_{dc}^2}$$

$$V_{dc} = \frac{2V_m}{\pi}$$

$$\text{Ripple factor (Theoretical)} = \sqrt{\left(\frac{V_{rms}}{V_{dc}}\right)^2 - 1} = 0.48$$

$$\text{Ripple Factor (Practical)} = \gamma = \frac{V_{ac}}{V_{dc}}$$

With RC Filter

$$\text{Ripple factor (Theoretical)} = \gamma = \frac{1}{4\sqrt{3}fCR}$$

$$V_{ac} = \frac{V_{r(p-p)}}{2\sqrt{3}} \qquad V_{dc} = V_m - \frac{V_{r(p-p)}}{2}$$

$$\text{Ripple Factor} = \gamma = \frac{V_{ac}}{V_{dc}}$$

$$\text{Percentage Regulation} = \left(\frac{V_{NL} - V_{FL}}{V_{FL}}\right) * 100\%$$

V_{NL} = No dc load voltage at the output side without connecting the load (Minimum current).

V_{FL} = Full-load dc voltage at the output side with load connected.

$$\text{Efficiency (in \%)} = \frac{dc\,output\,power}{ac\,output\,power} = \frac{p_{dc}}{p_{ac}} = \frac{V_{dc}^2/R_L}{V_{rms}^2/R_L}$$

$$\frac{V_{dc}^2}{V_{rms}^2} = \frac{\left[\frac{2V_m}{\pi}\right]^2}{\left[\frac{V_m}{\sqrt{2}}\right]^2} = \frac{8}{\pi^2} = 0.812 \equiv 81.2\%$$

The theoretical efficiency of a full-wave rectifier is 81.2%.

Model Graph of Full-Wave Centre-Tapped Rectifier
See Fig. 8.18.

Characteristics of a Full-Wave Rectifier

1. The peak voltage in the centre-tapped full-wave rectifier is only twice the peak voltage of that in the half-wave rectifier. This is because the secondary of the power transformer in the full-wave rectifier is centre-tapped and thus only half the source voltage goes to each diode.

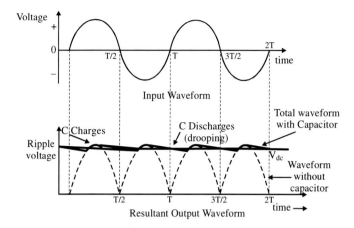

Fig. 8.18 Plot of voltage versus time circuit of full-wave centre-tapped rectifier for **a** input waveform, **b** output waveform

2. This rectifier requires a larger transformer for a given power output with two separate but identical secondary windings making this type of full-wave rectifying circuit costly as compared to the 'Full-Wave Bridge Rectifier'.

Precautions

1. During the experiment, we have to be careful not to exceed the ratings of the diode as this may damage the diode.
2. Switch **ON** the power supply only when we have checked the circuit connections as per the circuit diagram.

Procedure

1. Connections are to be made initially without capacitor (Fig. 8.17).
2. We now apply *ac* main voltage to the primary of the transformer and feed the rectified output voltage to the CRO for measuring the time period and amplitude of the waveform.
3. As a filter, now connect the capacitor in parallel to the load resistor and then note down the amplitude and time period of the waveform.
4. The amplitude and time period of the transformer secondary (input waveform) can be measured by CRO.
5. The input and output without filter and with filter voltage waveform may be plotted on a graph sheet with time in x-axis and voltage in y-axis as per the CRO screen scale.
6. Ripple factor can be calculated finally.

Formulae

$$\textbf{Peak} - \textbf{to} - \textbf{Peak Ripple Voltage (with filter)} = V_r(p - p) = \frac{V_{p(rect)}}{2fR_LC}$$

$$= 8 / \left(2 \times 43.5 \times 470 \times 470 \times 10^{-6} \right) = \textbf{0.416 V (Theoretical)}$$

where $V_{p(rect)}$ = Unfiltered peak (max.) rectified voltage.

$$\textbf{Peak rectified voltage} = V_{dc} = \left(1 - \frac{1}{4fR_LC} \right) V_{p(rect)}$$

$$= (1 - 0.026) \times 8 = \textbf{7.79 V (Theoretical)}$$

$$\textbf{Ripple Factor} = V_{r(pp)} / V_{dc} = \textbf{0.053}$$

Observation from CRO for the Given Circuit

Parameters voltage	Input	Output waveform (without filter)	Output waveform (with filter)	Ripple waveform (with filter)
Voltage amplitude	3.6 × 5 = 18 V	1.6 × 5 = 8 V	1.5 × 5 = 7.5 V	0.6 × 5 = 3 V
Time period (ms) (peak to peak)	4.6 × 5 = 23 ms	2.3 × 5 = 11.5 ms	2.3 × 5 = 11.5 ms	2.3 × 5 = 11.5 ms
Frequency (Hz)	43.0 Hz (mains)	86 Hz (Ripple)	86 Hz (Ripple)	86 Hz (Ripple)
Ripple voltage (V)		1.5 V	0.5 V	
Ripple factor (practical)		1.5/8 = 0.1875	0.5/7.5 = 0.067	

See Fig. 8.19.

Fig. 8.19 Actual graphical input and output waveforms as observed in CRO

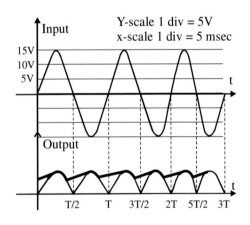

Result

The rectified output voltage in the given full-wave rectifier circuit was studied, and the values of ripple factor were found theoretically as **0.053** and as 0.066 through practical experiment here.

Viva Questions

Q.1   What is efficiency (η) of rectifier. Compare the efficiency of full-wave and half-wave rectifiers.

Q.2   Find the ripple voltage of a full-wave rectifier with filter capacitor of 50 μF connected to a 500 Ω load drawing 40 mA.

Q.3   The output frequency of a full-wave rectifier is _____ times the input frequency.

Q.4   What could be the approximate values of ripple factors of half-wave and full-wave rectifier.

Q.5   PIV in centre-tapped full-wave rectifier is $2V_m$, while in half-wave rectifier and full-wave rectifier, they have V_m. Explain for all the three.

8.9   Experiment 9: Full-Wave Bridge Rectifier

Aim
To study the working of a full-wave bridge rectifier with and without filter and to also measure its parameters.

Apparatus Requirement
Voltmeter (0–10 V),
Ammeter (0–50 mA) and (0–100 μA),
One transformer 6-0-6 V, 500 mA, 1 A rating,
One resistance 470 Ω, 10% tolerance, 1/2 W rating,
One capacitor 470 μF, and
Two diodes.

Introduction
In general, a rectifier converts a sinusoidal input waveform, i.e. *ac*, into a unidirectional waveform, i.e. *dc*, with non-zero average component. A full-wave rectifier converts an *ac* voltage into *dc* voltage using both half cycles of the wave as input *ac* voltage. In bridge rectifier, four diodes are connected to form a bridge as shown in Fig. 8.20, with load resistance ($R = R_L$) connected at the output parallel to the filter capacitor.

During the positive half cycle of the input *ac* voltage, diodes D1 and D3 conduct, whereas diodes D2 and D4 remain in the OFF state. As clear from Fig. 8.20, the conducting diodes will be in series with the load resistance ($R = R_L$) and therefore the load current flows through the load resistance. During the negative half cycle of the input *ac* voltage, diodes D2 and D4 conduct, whereas diodes D1 and D3 remain in the OFF state.

As seen in the circuit, the conducting diodes will be in series with the load resistance ($R = R_L$) and therefore load current flows through R in the same direction as in the previous half cycle. Thus, a bidirectional wave (*ac*) gets converted into a unidirectional wave (*dc*). Ripple factor is a measure of effectiveness of a rectifier circuit and is defined as a ratio of RMS (root mean square) value of *ac* component to the *dc* component in the rectifier output.

Theoretical Calculations
Ripple Factor

The ripple factor for a full-wave rectifier is given by

$$\gamma = \sqrt{\left(\frac{V_{rms}}{V_{dc}}\right)^2 - 1}$$

Circuit Diagram
See Fig. 8.20.

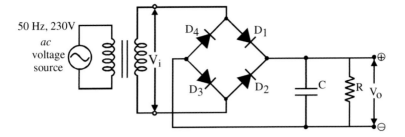

Fig. 8.20 Full-wave bridge rectifier circuit

The average voltage of the *dc* signal output available across the load resistance is $V_{dc} = 2V_m/\pi$ and the RMS value of the signal output at the load resistance is $V_{rms} = V_m/\sqrt{2}$. Therefore, **ripple factor** will become

$$\gamma = \sqrt{\left(\frac{\frac{V_m}{\sqrt{2}}}{\frac{2V_m}{\pi}}\right)^2 - 1} = \sqrt{\left(\frac{\pi}{8}\right)^2 - 1} = 0.48$$

Here, we see that the ripple factor can be lowered by increasing the value of the filter capacitor or by increasing the load.

Efficiency

Efficiency, η, is the ratio of *dc* output power to *ac* input power

$$\eta = \frac{dc\ output\ power}{ac\ input\ power} = \frac{P_{dc}}{P_{ac}}$$

$$= \frac{\frac{V_{dc}^2}{R_L}}{\frac{V_{rms}^2}{R_L}} = \frac{[2V_m/\pi]^2}{[V_m/\sqrt{2}]^2} = \frac{8}{\pi^2} = 0.812 = 81.2\%$$

The maximum efficiency of a full-wave rectifier is 81.2%.

Transformer Utilization Factor

Transformer utilization factor (TUF), which is defined as the ratio of power delivered to the load with *ac* rating of the transformer secondary; therefore,

TUF = *dc* power delivered to the load/*ac* rating of transformer secondary

TUF can be used to determine the rating of a transformer secondary, by considering the primary and the secondary windings separately. It has a value of 0.812 for an ideal transformer.

Characteristics of Full-Wave Bridge Rectifier

See Fig. 8.21.

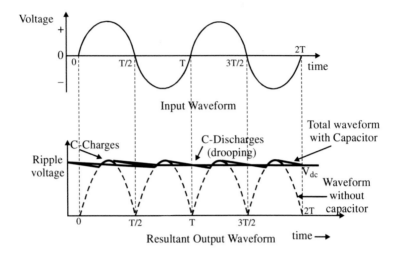

Fig. 8.21 Expected plot of voltage versus time of full-wave bridge rectifier for **a** input waveform, **b** output waveform

Precautions

1. While doing the experiment, do not exceed the ratings of the diode. This may damage the diode.
2. Connect CRO using probes properly as shown in the circuit diagram.
3. Do not switch **ON** the power supply unless you have checked the circuit connections as per the circuit diagram.

Procedure

1. Connections are given as per the circuit diagram without capacitor.
2. Apply *ac* main voltage to the primary of the transformer. Feed the rectified output voltage to the CRO and measure the time period and amplitude of the waveform.
3. Now connect the capacitor in parallel with load resistor and note down the amplitude and time period of the waveform.
4. Measure the amplitude and time period of the transformer secondary (input waveform) by connecting CRO.
5. Plot the input and output without filter and with filter waveform on a graph sheet.
6. Calculate the ripple factor.

Formulae
Peak-to-peak ripple voltage with capacitive load (filter),

$$V_{r(pp)} = \frac{V_{r(rect)}}{2fR_LC} = 0.22 \text{ V (Theoretical value)}$$

where $V_{p(rect)}$ = Unfiltered Peak Rectified Voltage = 9 V

Peak rectified voltage,

$$V_{dc} = \left(1 - \frac{1}{4}fR_LC\right)V_{p(rect)} = (1 - 0.012) \times 9 = \mathbf{8.8\,V}$$

$$Ripple\ Factor = V_{r(pp)}/V_{dc} = \mathbf{0.0252}\ (\text{for } R_L = 470\,\Omega, C = 470\,pf)$$

Observations

	Input waveform	Output waveform (without filter)	Ripple voltage (with filter)
Amplitude (max. to min.	3.8 × 5 = 19 V	1.5 × 6 = 9 V	1.4 × 0.2 = 0.28 V
Time period	4.4 × 5 ms = 22 ms	2.2 × 5 ms = 11 ms	2.2 × 5 ms = 11 ms
Frequency	45.45 Hz	90.9 Hz	90.9 Hz
Ripple factor			0.28/9 = **0.031**

Result

The rectified output voltage of full-wave bridge rectifier circuit was observed, and the calculated theoretical value of ripple factor is 0.0252, while in practical it was found to be a little higher at 0.031.

Curve for Full-Wave Bridge Rectifier

See Fig. 8.22.

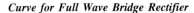

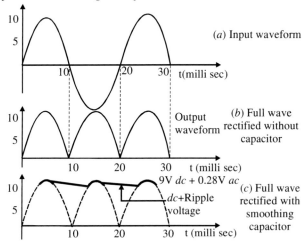

Fig. 8.22 Actual graphical input and output waveforms as observed in CRO

Viva Questions

Q.1   **Why the PIV of a diode full-wave bridge rectifier circuit is higher than the half-wave or bridge full-wave rectifiers. Explain why.**

Ans.   As in full-wave bridge, rectifier voltage of both the windings comes on the diode and therefore it has PIV $= 2V_m$.

Q.2   **Bridge rectifiers are preferred over full-wave rectifiers using centre-tapped transformer as (Tick the correct one)**

   (a) It uses four diodes
   (b) Its transformer is small and does not require centre-tapped
   (c) For the same output, it requires much smaller transformer
   (d) All of above

Ans.   (d) All of above

Q.3   **In all the rectifiers, the current in each diode flows for (Tick the correct one)**

   (a) Full cycle of the input signal
   (b) Half cycle
   (c) Less than half cycle
   (d) More than half cycle

Ans.   (b) Half cycle

Q.4   **Write the requirement of a filter in *dc* power supply.**

Ans.   To remove or reduce the ripple of the voltage waveform.

Q.5   **Define TUF. Write the TUF of half-wave, full-wave centre-tapped and bridge rectifier.**

Ans.   TUF is the short form of transformer utilization factor. TUF of half-wave rectifier = 0.287, full-wave centre-tapped rectifier = 0.693 and in full-wave bridge rectifier = 0.812.

Q.6   **With a 50 Hz *ac* signal given to a rectifier, the ripple frequency of the output voltage waveform for full bridge rectifier is**
   (a) 125 Hz (b) 50 Hz (c) 100 Hz (d) 200 Hz

Ans.   (c) 100 Hz

Q.7   **If 220 V *dc* voltage gets connected to a bridge rectifier in place of an *ac* source, the bridge rectifier will get damaged due to overloading of**
   (a) One diode (b) Two diodes (c) Three diodes (d) Full Bridge, i.e. all the four diodes

Ans.   (b) Two diodes

Q.8   **If our requirement is for low-voltage rectification which is a better choice:**

   (a) Two-diode full-wave rectifier will be better
   (b) Both bridge and full-wave rectifiers
   (c) Bridge rectifier or
   (d) None of them

Ans. (a) Two-diode full-wave rectifier will be better

Q.9   **As shown in Fig. 8.23 circuit find (a) the *dc* output voltage, (b) the *dc* load current, (c) the RMS value of the load current, (d) the *dc* output power, (e) the *ac* power, (f) the efficiency of rectifier, (g) peak inverse voltage of each diode, and (h) output voltage frequency. Assume all diodes are ideal.**

Ans. (a) 49.50 V (b) 0.0997 A (c) 1.1 A (d) 4.918 W (e) 96.6 W (f) 5.07% (g) 77.78 V (h) 99 Hz

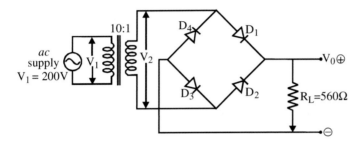

Fig. 8.23 Bridge rectifier circuit

8.10   Experiment 10: LED Colour Characteristics

Aim
To study the general colour output characteristics of light-emitting diode (LED).

Apparatus
Regulated power supply (0–10 V),
Resistor 330 Ω,
Ammeters (0–100 mA),
Voltmeter (0–30 V),
LED: 1 red colour, 1 yellow colour and 1 green colour, and
Breadboard and connecting wires.

Introduction
An LED emits light when electric current passes through them only in forward bias. It is a normal *pn* junction diode except that the semiconductor material is GaAs or InP, causing the colour of the light. (For details, see Chap. 4.)

When it is forward biased, the mobile charges move, which causes holes to move from *p* to *n* region and electrons flow from *n* to *p region*. At the junction, the opposite charge carriers recombine with each other and released the energy in the form of light and thus LED emits light under forward-biased condition. Under reverse-biased condition, there is very little recombination due to minority carriers current; as a result, there is nearly zero emission of light. In *Si* diode also such emission takes place but in infrared region, which is not visible.

Connecting and Soldering
LEDs must be connected as per the terminal anode, cathode for which indicators on the packaging may be (i) anode labelled **A** or + sign and **K** or −ve for cathode (yes, it really is K, not c, for cathode) or (ii) cathode has shorter lead than anode or (iii) on the cathode side there may be a slight flat region on the body of the round LEDs. Normally, all the time the anode lead (+ve) *p* side of the LED is longer.

LEDs can get damaged by heat when soldering; therefore, we need to be very slow; otherwise, no special precautions are needed for soldering most LEDs. In Fig. 8.24b, a two-colour LED is shown with two anodes with common cathode (*k*).

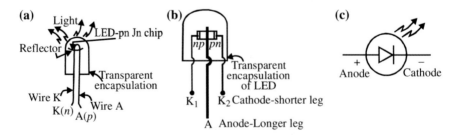

Fig. 8.24 a Active transparent encapsulated single LED **b** bicolour LED with *np* diode made of one material and *pn* diode of another material, **c** symbol of LED

Here, the two diodes are of different materials. If an *ac* square wave of very low frequency say *f* < 10 c/s is applied across a1 and a2 with k open, then the change of colour of LED with that frequency will be visible to our eyes.

Testing an LED
We should not connect an LED directly to a battery or power supply. It will get burnt almost instantly because too much current will pass through, and therefore LEDs must have a resistor in series to limit the current to a safe value. For quick testing purposes, a 1 kΩ resistor is suitable for most LEDs with a supply voltage of 12 V or less.

Colours of LEDs
LEDs are available normally in all colours, e.g. red, orange, amber, yellow, green, blue and white. Blue and white LEDs are much more expensive than the other colours due to special material used (see Table 4.1). The colour of an LED is determined by the semiconductor material used (see Table 4.1), not by the colouring of the 'package' (the plastic body, which may be diffused (milky) or clear (often described as 'glass clear').

The coloured packages are also available as diffused (the standard type) or transparent. As well as a variety of colours, sizes and shapes, LEDs also vary in their viewing angle. This tells you how much the beam of light spreads out. Standard LEDs have a viewing angle of 60° but others have a narrow beam of 30° or less.

Calculating Series Resistor Value for an LED
An LED must have a resistor connected in series (as shown in Fig. 8.25a) for limiting the current through the LED; otherwise, it will get burnt almost instantly. The minimum resistor value, R, is given by

$R = (V_S - V_L)/I$
V_S = supply voltage
V_L = LED voltage (usually 2 V, but 4 V for blue and white LEDs)
I = LED current (e.g. 20 mA), this must be less than the maximum allowed as per specification of the LED being used.

The calculated value may not be available all the time; therefore, we choose the nearest standard resistor value, which is greater than the resistor value to reduce the current (to increase LED and battery life); however, this will make the LED less bright.

Fig. 8.25 LED connected with series resistor

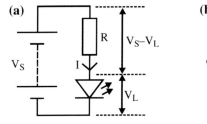

As an example, if the supply voltage is $V_S = 9$ V, and we have a red LED ($V_L = 2$ V), requiring a current $I = 20$ mA $= 0.020$ A, $R = (9\,V - 2\,V)/0.02\,A = 350\,\Omega$, therefore we choose 390 Ω (the nearest standard value which is greater).

Connecting LEDs in Series

We may wish to have several LEDs on at the same time by connecting them in series, which prolongs battery life by lighting several LEDs with the same current as just one LED (Fig. 8.26).

For the LEDs connected in series, they should all be of the same type. The power supply must have voltage to provide about 2 V for each LED (4 V for blue and white) plus at least another 2 V for the resistor. For working out a value for the resistor, we must add up all the LED voltages and use this for V_L.

As an example, three LEDs a red, a green and a yellow in series will need a voltage (V_S) of at least 3×2 V + 2 V (for resistor) = 8 V; therefore, **9 V battery** will suffice. Regarding resistor value, for the supply voltage of 9 V and the current (I) requirement of 15 mA $= 0.015$ A, the resistor $R = (V_S - V_L)/I = (9 - 6)/0.015 = 3/0.015 = 200\,\Omega$; therefore, we may choose the value of R as 220 Ω (the nearest standard value which is greater).

We have to avoid connecting LEDs in parallel (Fig. 8.26b). Connecting several LEDs in parallel with just one resistor shared between them is generally not a good because if the LEDs require slightly different voltages only the voltagelowest LED will light and it also gets damaged by the larger current flowing through it. Although identical LEDs can be connected in parallel with one resistor, this rarely offers any useful benefit because resistors are very cheap.

Advantages of LED

1. It requires less complex circuitry.
2. Fabrication is cheap with high yield.
3. Life longevity is as large as 100,000 h, and because of this, LED bulbs for lighting are becoming popular.

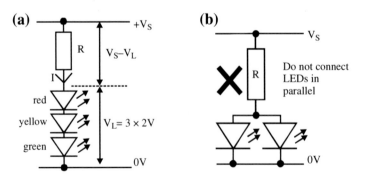

Fig. 8.26 a Three LEDs connected in series, **b** LEDs connected in parallel to be avoided

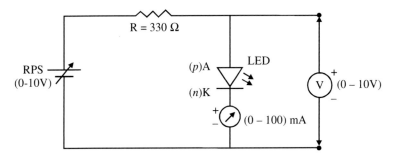

Fig. 8.27 Circuit for forward bias characteristics

Desired Characteristics

1. It is hard to external radiation.
2. It has very high quantum efficiency.
3. It also has fast emission response time.

Basic LED Configuration

1. Edge emitter: light emitted from the edge from the junction.
2. Surface emitter: light emitted from the surface through the shallow transparent *p* layer.

Circuit Diagram: Forward Bias
See Fig. 8.27.

Procedure

1. Connection to be made as per the circuit diagram as shown in Fig. 8.27.
2. Now vary the input voltages of the regulated power supply (RPS) and we note down the corresponding current in the LED and the voltage across it.
3. Repeat these for reverse bias condition and we note down the corresponding voltages and currents.
4. Finally, we plot the graph between voltage and current of LED in forward bias and reverse bias as per the table given below. We note that with higher element, intensity of light increases.

Observations

LED colour: red		LED colour: yellow		LED colour: blue	
Volt (V)	I (mA)	Volt (V)	I (mA)	Volt (V)	I (mA)
0.6	0	0.5	0	0.7	0
0.9	0	0.9	0	1.4	0
1.4	0	1.6	0	1.7	0
1.6	2	1.8	2	1.8	2

(continued)

(continued)

LED colour: red		LED colour: yellow		LED colour: blue	
Volt (V)	I (mA)	Volt (V)	I (mA)	Volt (V)	I (mA)
1.8	3	2.0	3	1.9	4
1.9	4	2.1	4	2.1	5
1.9	6	2.2	6	2.2	7
2.0	7	2.2	7	2.3	9
2.0	9	2.2	9	2.5	11
2.0	10	2.3	10	2.6	13
2.0	11	2.3	11	2.6	14
2.1	13	2.3	13	2.6	16
2.1	14	2.3	14	2.6	17
2.2	16	2.3	16	2.7	18
2.2	17	2.4	17	2.8	19
2.2	19	2.5	19	2.9	20

Maximum allowed element, gives maximum light intensity

Observation I-V Curve
See Fig. 8.28.

Result
This way, the I-V characteristic of LED was studied and noted that voltage need of blue LED is highest and lowest in red LED, as semiconductor material is different.

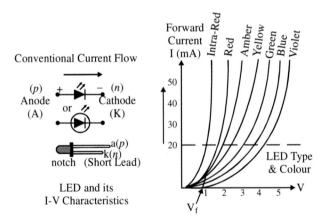

Fig. 8.28 LED symbol and I-V forward characteristics

Viva Questions

Q.1   **How LED is different from normal *pn* junction diode?**

Ans. If LED is similar to normal *pn* junction diode except for the fact that the basic semiconductor material is GaAs or InP, which is responsible for the colour of light of that LED, *pn* junction of Si also emits radiation but in IR range.

Q.2   **Define wavelength.**

Ans. Wavelength (λ) of light is defined as the length of one complete wave, as the distance between two adjacent peak voltage positions of time or two adjacent droughts.

Q.3   **When LEDs are connected in series or parallel, what will be the power needed?**

Ans. LEDs are connected in series; all of them glow at the same time as the same current flow through them, giving equal intensities. LEDs should not be connected in parallel as different LEDs may require different voltages to operate; hence, as a result some LEDs requiring low voltage may get destroyed eventually.

Q.4   **Write the advantages of LED over laser diode?**

Ans. Fabrication as well as operation of LED is less complex than laser diode and can be fabricated with ease than laser diode. As a result, LEDs are much cheaper.

Q.5   **What are the desired characteristics of LED?**

Ans. (a) It is hard/tolerant to external radiation
      (b) It has fast emission response time
      (c) It has high quantum efficiency

8.11 Experiment 11: Transistor in Common Emitter Configurations

Aim

To study and plot the input–output characteristics of a bipolar junction transistor in common emitter configuration and also compute the two h-parameters (i.e. h_{ie} and h_{oe}).

Apparatus Requirement

Regulated power supply (0–30 V),
Resistor 1 kΩ,
1 *npn* transistor BC147,
Ammeters (0–10 mA, 0–500 mA),
Voltmeter (0–1 V, 0–30 V),
LED: 1 red colour, 1 yellow colour and 1 green colour,
Breadboard and connecting wires.

Introduction

A bipolar junction transistor (BJT) is a three-terminal (emitter, base, collector) semiconductor device, normally used for signal amplification and as a digital switch. The pin assignment is shown in Fig. 8.29. There are two types of bipolar transistors, namely, *npn* and *pnp*, both consisting of two *pn* junctions, namely, emitter junction and collector junction.

The common emitter (CE) configuration (Fig. 8.29) has the input, applied between base and emitter; and the output is taken across collector and emitter. Here, emitter is common to both input and output and therefore the name common emitter (CE) configuration.

Input characteristic measurements are between the input current and input voltage, taking output voltage as parameter. It is plotted between V_{BE} and I_B at constant V_{CE} in *CE* configuration.

Output characteristic measurements are between the output voltage and output current, taking input current as parameter. It is plotted between V_{CE} and I_C at constant I_B.

Pin Assignment

See Fig. 8.29.

Circuit Diagram

See Fig. 8.30.

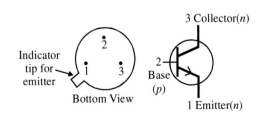

Fig. 8.29 Pin assignment of *npn*–BJT. Here tip or dot on package indicates Emitter(1), then at the bottom count clockwise Base(2) and Collector(3)

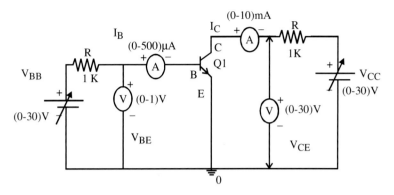

Fig. 8.30 Biasing circuit of *npn*–BJT with CE configuration

Precautions

1. Doing the experiment, we have to be careful not to exceed the ratings of the transistor, as this may damage it.
2. Voltmeter and ammeter are also to be connected in correct polarities as shown in the circuit diagram.
3. Before switching **ON** the power supply, check the circuit connections as per the diagram (Fig. 8.30).
4. Also be sure while connecting the emitter, base and collector terminals of the transistor.

Procedure
Input Characteristics

1. Connect the *npn*–BJT transistor as per circuit diagram in Fig. 8.30.
2. Maintain the output voltage $V_{CE} = 0$ V by varying V_{CC} for the first set of readings (see below Table 8.4).
3. Now vary V_{BB} gradually from 1 V and note down both input base current I_B and base–emitter voltage (V_{BE}) to get the first set of readings.
4. For the second set of readings, maintain the output voltage $V_{CE} = 5$ V and repeat step 3 (see Table 8.4).
5. Plot $I_B - V_{BE}$ graph for $V_{CE} = 0$ V and $V_{CE} = 5$ V.

Output Characteristics

1. Connections of the components to be as per Fig. 8.30.
2. Now, we maintain $I_B = 20$ μA constant by changing V_{BB}, for the first set of readings (see Table 8.4).
3. Now vary V_{CC} gradually, and note down both outputs, i.e. collector current I_C and collector–emitter voltage (V_{CE}) to get the first set of readings.
4. For second set and third set of readings, change $I_B = 30$ μA and 60 μA by regulating V_{BB} (see Table 8.5).

5. Plot I_B–V_{BE} graph for three sets of I_B = 20 μA, 30 μA and 60 μA.

Observations
See Tables 8.4 and 8.5.

Input and Output I-V Characteristics of CE Transistor Configuration Graph (Expected and Experimental)
See Fig. 8.31a–c

Result
Thus, we have studied the input and output characteristics of BJT in CE configuration and verified the graph by plotting it. Now, the input impedance and output admittance, and hence, the corresponding two h-parameters (h_{ie} and h_{oe}) are as per slopes from Fig. 8.31b, c as follows:

(a) The input impedance (h_{ie}) = $\Delta V_{BE}/\Delta I_B$, (at Constant V_{CE} = 0 V) = 0.1/80 = 1.25×10^{-3} Ω = Z_{in} (**Point P**).
(b) The input impedance (h_{ie}) = $\Delta V_{BE}/\Delta I_B$, (at Constant V_{CE} = 5 V) = 0.1/86 = 1.163×10^{-3} Ω = Z_{in} (**Point Q**).
(c) The output admittance (h_{oe}) = $\Delta I_C/\Delta V_{EC}$ (at constant I_B = 30 μA) = 0.023 **milli mhos**; Z_{out} = 100 kΩ (**Point R**).
(d) The output admittance (h_{oe}) = $\Delta I_C/\Delta V_{EC}$ (at constant I_B = 60 μA) = 0.166 **milli mhos**; Z_{out} = 10 kΩ (**Point S**).

Table 8.4 Input characteristics (by readings)

V_{CE} = 0 V		V_{CE} = 5 V	
V_{BE} (V)	I_B (μA)	V_{BE} (V)	I_B (μA)
0.4	0	0.4	0
0.5	5	0.5	0
0.6	30	0.6	10
0.7	180	0.7	80
0.8	400	0.8	320

Table 8.5 Output characteristics (by readings)

I_B = 30 μA		I_B = 60 μA	
V_{CE} (V)	I_C (mA)	V_{BE} (V)	I_C (mA)
1	1.2	1	3.5
2	1.2	2	3.6
3	1.2	3	3.7
4	1.2	4	3.8
5	1.2	5	3.9
6	1.2	6	4.0
7	1.2	7	4.1
8	1.2	8	4.2

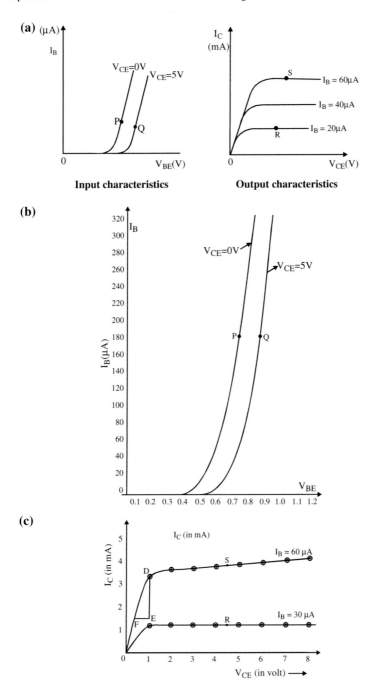

Fig. 8.31 a Expected input and output characteristics. **b** Experimental input characteristics. **c** Experimental output characteristics

Thus, we have proved that in CE configuration, the input impedance (Z_{in}) is low but the output impedance (Z_{out}) is quite large. The computed h-parameters h_{ie} and h_{oe} are also given above.

Viva Questions

Q.1 Why is the base layer in a transistor kept very thin?

Q.2 The junction capacitance around collector to base junction is much lower than that around base to emitter junction. Why?

Q.3 What is the switching action in a transistor?

Q.4 What is base width modulation?

Q.5 Design another *npn* common emitter transistor circuit with proper biasing to work in active region of the transistor.

Q.6 A certain region of the output characteristics of a transistor is used as an amplifier. Which one and why?

Q.7 Silicon transistor is more commonly used compared to germanium transistor or gallium arsenide transistor. Why?

Q.8 What is the switching action in a transistor? What is the switching speed in a normal *npn* transistor?

Q.9 If we choose a transistor for a particular application, say oscillator, what will be the basis or parameters for taking a decision?

Q.10 What is the collector to emitter terminal voltage when the transistor is in (i) saturation region, (ii) cut-off region, (iii) active region.

8.12   Experiment 12: JFET Characteristics

Aim
To study and plot drain and transfer characteristics of a FET and compute the related parameters

Apparatus Requirement
Regulated power supply (0–30 V), 2 A,
Resistor 1 kΩ,
1 JFET BFW10,
Ammeters (0–20 mA),
Voltmeter (0–1 V, 0–20 V), and
Breadboard and connecting wires.

Introduction
A *n*-channel field-effect transistor (FET) is made of *n*-type material called the substrate with a *p*-type (the gate) diffused into it in a small region. Two metal controls are made in the non-diffused region across the gate, called drain and source. This region between source and drain is called channel. With a positive voltage on the drain, with respect to the source, electron current flows from source to drain through the channel. If the gate is made negative (reverse biased) with respect to the source, the *pn* junction gate gets reverse biased, creating a charge depleted region of decreasing width from source to drain. The electrostatic field created minimizes the channel width due to depletion layer formation which reduces the drain current. (**Refer to Theory; Article 3.6**, Fig. 3.23 and Fig. 8.32.)

Circuit Diagram
See Fig. 8.33.

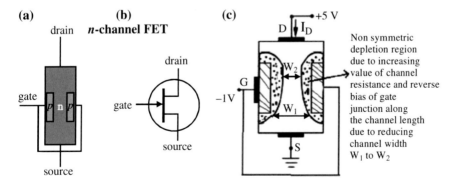

Fig. 8.32 *n*-channel JFET **a** two-dimensional structure, **b** its symbol, **c** channelling processing

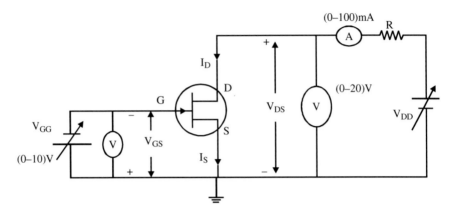

Fig. 8.33 Circuit diagram of common source (CS) *n*-channel JFET with proper biasing

Precautions
1. While doing the experiment, we have to careful not to exceed the ratings of the FET as this may damage the FET.
2. Voltmeter and ammeter polarities should be noted before connecting them as shown in the circuit diagram.
3. Switch ON the power supply after being checked the circuit connections as per the circuit diagram.
4. While selecting the source, drain and gate terminals of the FET, refer Fig. 8.34.

Pin Assignment of *n*-channel JFET (BFW10)
See Fig. 8.34.

Characteristics of JFET
1. Transconductance (g_m) of JFET at zero gate–source voltage is generally in the range of 0.1–10 mA/V, since drain current is proportional to g_m.
2. The gate leakage current of JFET is normally in the range of 100 µA–10 nA, i.e. it consumes a little power, whereas in MOSFET, it is 100 nA–10 pA and therefore consumes very little power.
3. Due to greater susceptibility of damage to JFET even due to some static charges, we have to be very careful.

Fig. 8.34 Pin assignment of *n*-channel JFET (BFW10). Tip or dot on the package indicates Source(1), then at the bottom count clockwise Drain(2), Gate(3) and Substrate (4)

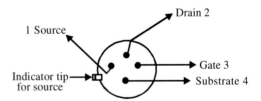

Observations

Drain characteristics				Transfer characteristics			
$V_{GS} = 0$ V		$V_{GS} = -2$ V		$V_{DS} = 2$ V		$V_{DS} = 4$ V	
V_{DS} (V)	I_D (mA)	V_{DS} (V)	I_D (mA)	V_{GS} (V)	I_D (mA)	V_{GS} (V)	I_D (mA)
2	11.2	2	3.7	0	8	0	9.5
4	13.6	4	4.8	1	6	1	6.6
6	14.1	6	6.1	2	3	2	5.3
8	14.2	8	7.2	3	0.5	3	2.5
10	14.2	10	7.21	4	0	4	0.3
12	14.2	12	7.21	5	0	5	0

Model Graph

See Fig. 8.35 (Expected).

See Fig. 8.36 (Experimental).

Graph (Instructions)

1. Plot the drain characteristics by taking V_{DS} on X-axis and I_D on Y-axis at constant V_{GS}.
2. Plot the transfer characteristics by taking V_{GS} on X-axis and I_D on Y-axis at constant V_{DS}.

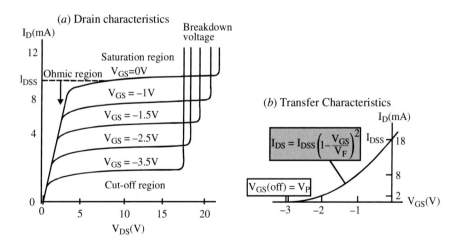

Fig. 8.35 Expected **a** drain characteristics, **b** transfer characteristics

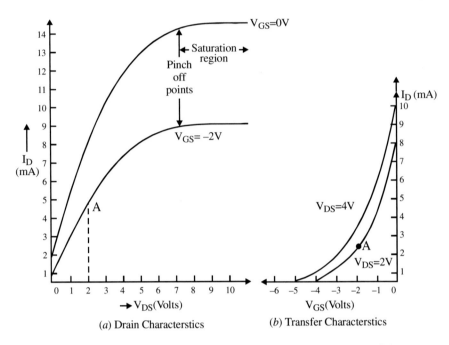

Fig. 8.36 Experimental (CRO) results, **a** drain characteristics, **b** transfer characteristics

Calculations from Graph

Drain Resistance (r_d)
When the JFET is operating in pinch-off or saturation region, $[r_d = (\Delta V_{DS}/\Delta I_D)]$, the output resistance of JFET (r_d), the ratio of small change in drain-to-source voltage (ΔV_{DS}) to the corresponding change in Drain current (ΔI_D) for a constant gate-to-source voltage (V_{GS}).

Transconductance (g_m)
Ratio of small change in drain current (ΔI_D) to the corresponding change in gate-to-source voltage (ΔV_{GS}) for a constant V_{DS} is called transconductance ($g_m = \Delta I'_D \Delta V_{GS}$ at constant V_{DS}). The word 'trans' signifies the input-to-output relation, i.e. from transfer characteristics. The value of g_m is expressed in mhos.

Amplification Factor (μ)
It is the ratio of small change in drain-to-source voltage (ΔV_{DS}) to the corresponding change in gate-to-source voltage (ΔV_{GS}), with constant drain current.

$$\mu = \Delta V_{DS}/\Delta V_{GS}$$
$$\mu = (\Delta V_{DS}/\Delta I_D) \times (\Delta I_D/\Delta V_{GS})$$
$$\mu = r_d \times g_m$$

Voltage amplification (μ) in JFET is not very high ($\mu \approx 5$–20) but it has the advantage of high input impedances, thermal and radiation stability, low power mode, etc. Currentgain in BJT is quite high ($\beta \approx 100$) along with voltage gain also. Hence, power gain in BJT is quite high.

Result

This way, the drain and transfer characteristics of a JFET were studied and verified the characteristics. The parameters calculated from slopes of various curves are as follows (e.g. Fig. 8.36):

(i)   Drain resistance (r_d) at ($V_{GS} = 0$) = 0.769 kΩ.

(ii)   Drain resistance (r_d) at ($V_{GS} = -2$ V) = 1.5385 kΩ.

(iii)   Transconductance (g_m) at ($V_{DS} = 2$ V) = 3 m mhos. ($V_{GS} = 0$)

(iv)   Transconductance (g_m) at ($V_{DS} = 2$ V) = 1.3 m mhos. ($V_{GS} = -2$ V)

(v)   Amplification factor (μ) = 0.769 kΩ × 3 m mho = 2.307. ($V_{GS} = 0$ V)

(vi)   Amplification factor (μ) = 1.5385 kΩ × 1.3 m mho = 2.00. ($V_{GS} = -2$ V)

Viva Questions

Q.1 **Why current gain (β) is an important parameter in BJT, whereas conductance is important parameter in FET?**

Ans. A bipolar junction transistor is a current-controlled device. Its output characteristics (μ) are controlled by base current; therefore, output characteristics of I_C–V_{CE} are more important.

     In FET, the output characteristics are controlled by gate voltage; therefore, transfer characteristics are now important and so are the transconductance, more important. The currentgain is important parameter for BJT, whereas for FET, voltage gain is important.

Q.2 **In JFET, the drain current remains physically constant above pinch-off voltage. Why?**

Ans. Beyond pinch-off voltage, the channel becomes very small. So, the depletion layer almost touches each other, leaving very small region with carriers for drain current I_D. Therefore, the small drain current passes through small width between depletion layers and increase of current is very small with increase in drain voltage.

Q.3 **How can avalanche breakdown be avoided in FET.**

Ans. Breakdown of a device is caused by very high electric field. Hence, the junction region of a FET should be doped lightly to reduce the electric field.

Q.4 **State the reason for the depletion region around the gate in the channel area, to be of non-symmetrical nature.**

Ans. In n-channel depletion mode, a −ve gate-to-source voltage forms a depletion region with no mobile carrier. The drain–source voltage will cause I_D to flow. The voltage if measured from source to any point (V_x) between source to drain will come out to increase from zero near source linearly to full value, V_D at drain; therefore, the depletion width also keeps on increasing, leading to non-symmetrical structure.

Q.5 **How hole flow through p-channel in FET?**

Ans. In p-type semiconductor, the gate is formed due to n-type doping. The hole will be the majority carrier in the channel due to which they are current carriers.

Appendix: Constants, Units and Symbols

See Tables A.1, A.2, A.3, A.4, A.5, A.6, A.7, A.8 and A.9.

Table A.1 Physical constants

Velocity of light = c = 2.998×10^8 m/s
Charge of electron = e = 1.602×10^{-19} C
Mass of electron = m = 9.107×10^{-31} kg
Electron charge-to-mass ratio (e/m) = 1.76×10^{11} C/kg
Boltzmann constant k = 1.380×10^{-23} J/K
Planck's constant h = 6.547×10^{-34} J s
Permittivity of free space = ϵ_0 = 8.854×10^{-12} F/m
Permeability of free space = μ_0 = $4\pi \times 10^{-7}$ H/m
Impedance of free space Z_0 = 376.7 = $120\pi\,\Omega$
Thermal voltage (kT) = 0.0259 eV; kT/e = 0.0259 V

© Springer Nature Singapore Pte Ltd. 2020
S. S. Srikant and P. K. Chaturvedi, *Basic Electronics Engineering*,
https://doi.org/10.1007/978-981-13-7414-2

Table A.2 Conversion factor between units

1 Angstrom	10^{-4} micron = 10^{-8} cm
1 foot	0.305 m
1 gauss	10^{-4} tesla
1 inch	2.54 cm
1 kg	2.2 lb
1 lb	453.6 gm
1 micron	10^{-6} m = 10^{-4} cm
1 cm	393.7 mils
1 mil	10^{-3} inch = 2.54×10^{-3} cm = 25.4 micron
1 mile	1.61 km
1 dB	0.115 nepers
1 neper	8.686 dB

Table A.3 Free carriers densities in metal and semiconductors (No. of carriers/cc)

Metals	Semiconductors (intrinsic)	Semiconductors (extrinsic doping range)
Cu: 8.47×10^{22}	Si: 1.5×10^{10}	Si: 10^{11}–10^{20}
Ag: 5.86×10^{22}	GaAs: 1.8×10^{10}	GaAs: 10^7–10^{21}
Au: 5.90×10^{22}	Ge: 2.3×10^{13}	Ge: 10^{14}–10^{25}
Al: 18.1×10^{22}	GaN: 10^{13}	GaN: 10^{14}–10^{20} (i.e. *intrinsic to metallic*)

Table A.4 Prefixes of units

Quantity	Symbol	Unit
exa	E	10^{18}
peta	P	10^{15}
tera	T	10^{12}
giga	G	10^9
mega	M	10^6
kilo	k	10^3
hecto	h	10^2
deka	da	10
deci	d	10^{-1}
centi	c	10^{-2}
milli	m	10^{-3}
micro	μ	10^{-6}
nano	n	10^{-9}
pico	p	10^{-12}
femto	f	10^{-15}
atto	a	10^{-18}

Table A.5 Properties of silicon, gallium arsenide and germanium (T = 300 K)

Property	Si	GaAs	Ge
Atom/CC	5.0×10^{22}	4.42×10^{22}	4.42×10^{22}
Atomic weight	28.09	144.63	72.60
Crystal structure type	Diamond	Zincblede	Diamond
Density (g/cm^{-3})	2.33	5.32	5.33
Breakdown field $|V|$ (cm)	$\approx 3 \times 10^5$	$\approx 4 \times 10^5$	$\approx 10^5$
Lattice constant (Å unit)	5.43	5.65	5.65
Melting point (°C)	1415	1238	937
Dielectric constant (ε_r)	11.7	13.1	16.0
Band gap energy (eV)	1.12	1.42	0.66
Electron affinity, χ (volts)	4.01	4.07	4.13
Resistivity, ρ (Ω cm)	2×10^5	2×10^9	4.7
Effective density of states in conduction band, N_c (cm^{-3})	2.8×10^{19}	4.7×10^{19}	1.04×10^{19}
Effective density of states in valence band, N_v (cm^{-3})	1.04×10^{19}	7.0×10^{18}	6.0×10^{18}
Intrinsic carrier concentration (cm^{-3}) (n_i)	1.5×10^{10}	1.8×10^8	2.3×10^{13}
Electron, mobility (μ_n) (cm^2/V s)	1350	8500	3900
Hole mobility, (μ_p) cm^2/V s	480	400	1900

Table A.6 Conductivity of conductors and insulators (σ) Mhos/metre

Conductor/semicond.	σ	Insulator	σ
Silver	7.17×10^7	Quartz	10^{-17}
Copper	5.80×10^7	Polystyrene	10^{-16}
Gold	4.10×10^7	Rubber (hard)	10^{-15}
Aluminium	3.82×10^7	Teflon	10^{-24}
Tungsten	4.82×10^7	Mica	10^{-14}
Iron	1.03×10^7	Porcelain	10^{-13}
Solder	0.70×10^7	Glass	10^{-12}
Steel (stainless)	0.11×10^7	Sand (dry)	2×10^{-4}
Nichrome	0.10×10^7	Air	5×10^{-15}
Graphite	2×10^7 to 10^5	Clay	10^{-4}
Silicon	1.56×10^{-3}	Rubber (hard)	10^{-15}
GaAs	10^{-8} to 10^3	Water (deionised)	10^{-6}
Ge	2.17	Water (distilled)	2×10^{-4}
		Water (fresh)	10^{-3}
		Water (Sea)	3–6
		Ferrite (typical)	10^{-2}

Table A.7 Dielectric constant, i.e. relative permittivity (ε_r)

Material	ε_r	Material	ε_r
Air	1	Sand (dry)	4
Alcohol (ethyl)	25	Snow	3.3
Bakelite	4.8	Soil (dry)	2.8
Glass	4–7	Styrofoam	1.03
Mica (ruby)	5.4	Teflon	2.1
Nylon	4	Water (distilled)	80
Paper	2–4	Water (sea)	20
Polyethylene	2.25	Wood (dry)	1.5–4
Porcelain (dry process)	6	Ground (wet)	5–30
Quartz (fused)	3.80	Ground (dry)	2–5
Rubber	2.5–4	Water (fresh)	80

Table A.8 Relative permeability μ_r

Diamagnetic material	μ_r	Ferromagnetic material	μ_r
Bismuth	0.99999	Nickel	50
Paraffic	0.99999	Cast iron	60
Wood	0.99999	Cobalt	60
Silver	0.99999	Machine steel	300
Paramagnetic material	μ_r	Ferrite steel	1,000
Aluminium	1.00000065	Transformer iron	3,000
Beryllium	1.00000079	Iron (Pure)	4,000
Mumetal	20,000	Supermalloy	100,000

Table A.9 List of symbols and units

Symbols	Description	Units
n	Electron concentration	cm^{-3}
p	Hole concentration	cm^{-3}
N_D	Donor concentration	cm^{-3}
N_A	Acceptor concentration	cm^{-3}
N_n	Majority carriers as electrons in n-type semiconductor	cm^{-3}
N_p	Minority carriers as holes in n-type semiconductor	cm^{-3}
P_p	Majority carriers as holes in p-type semiconductor	cm^{-3}
P_n	Minority carriers as holes in n-type semiconductor	cm^{-3}
E_g	Forbidden energy gap	eV
E_C	Energy level below conduction band	eV
E_V	Energy band above the valence band	eV
E_F	Fermi level	eV
E_i	Intrinsic energy level	eV

(continued)

Table A.9 (continued)

Symbols	Description	Units
$f(E)$	Fermi–Dirac distribution formulae	–
C	Capacitance	Farad
L	Inductance	Henry
c or v	Velocity of light in air or vacuum	cm/s
I	Current	Ampere or A
L	Channel length	cm or μm
l	Length	m or cm
A	Area	m^2 or cm^2
V	Volume	m^3 or cm^3
J	Current density	A/cm^2
J_n	Electron current density	A/cm^2
J_n	Hole current density	A/cm^2
h	Planck constant	Joule sec
$h\nu$	Photon energy	eV
I_C	Collector current	mA
I_{CEO}	Leakage current flowing between collector and emitter with base open	μA or nA
I_{CBO}	Leakage current in the reverse-biased CB junction with emitter open	μA or nA
I_E	Emitter current	mA
I_B	Base current	μA
I_O	Reverse saturation current of the diode	μA
I_D	Drain current	mA
m_e^*	Effective mass of electron	kg
m_h^*	Effective mass of hole	kg
Q	Charge	Coulomb
α	Common base current gain of BJT	–
β	Common emitter current gain of BJT	–
γ	Common collector current gain of BJT	–
V_b	Built-in voltage	volt
V_{th}	Threshold	volt
t	Time	sec
T	Temperature	°K or °C
Ψ	Flux linkage	weber = volt. sec
R_L	Load resistance	ohm (Ω)
ϕ_m	Semiconductor work function	cm^{-3}
φ	Electric potential	V
ϕ_m	Semiconductor work function	cm^{-3}
φ	Electric potential	V
W	Power	watt = Joule/sec
f	Frequency	Hz = cycle per sec

(continued)

Table A.9 (continued)

Symbols	Description	Units
ω	Angular frequency	rev/sec
k	Boltzmann constant	Joule per Kelvin (JK)
m_a	Modulating index of AM	–
m_f	Modulating index of FM	–
V_{sat}	Saturation voltage	V
v_d	Drift velocity	m/s
μ_n	Mobility of electron	cm^2/Vs
μ_p	Mobility of hole	cm^2/Vs
ρ	Resistivity	Ω m
σ	Conductivity	Siemens per metres (S/m) = (Ωm)
τ_p	Hole lifetime	s
τ_n	Electron lifetime	s
F	Force	N
E	Electric field	V/m

Note 1 Tesla = 1 weber/m^2 = 10^4 G = 3 × 10^{-6} ESU

1 Å (Angstrom) = 10^{-10} m

1 μm (micron) = 10^{-6} m = 10,000 Å

Bibliography

Gupta N (2010) Electronics engineering (basic electronics). Dhanpat Rai and Sons
Jacob S, Halkias C (1967) Electronic devices and circuits. McGraw-Hill
Mandal SK (2012) Basic electronics. McGraw-Hill
Millman J, Grabel J (2001) Microelectronics. Tata McGraw Hill
Muthusubramanian R, Salivahanan R (2012) Basic electrical and electronics engineering. Tata McGraw-Hill Education
Sharma S (2013) A textbook of electronics engineering, 2013 ed. S.K. Kataria & Sons
Sze SM (1969) Physics of semiconductor devices. Wiley-Interscience

© Springer Nature Singapore Pte Ltd. 2020   373
S. S. Srikant and P. K. Chaturvedi, *Basic Electronics Engineering*,
https://doi.org/10.1007/978-981-13-7414-2

Subject Index

© Springer Nature Singapore Pte Ltd. 2020
S. S. Srikant and P. K. Chaturvedi, *Basic Electronics Engineering*,
https://doi.org/10.1007/978-981-13-7414-2

Printed in the United States
by Baker & Taylor Publisher Services